MiGS OVER NEVADA

US MiG EXPLOITATION PROJECTS

Author with MiG-17 at Groom Lake Facility - 1969

By:

Thornton D. "TD" Barnes

Former Member Area 51 Special Projects Team
Codename: Thunder

Copyright © 2014 Thornton D. Barnes

All rights reserved.

ISBN-13:978-1499551655

ISBN-10:1499551657

Dedication

I dedicate this book about the Soviet MiG exploitation projects in Nevada to all who encountered in combat the Mikoyan-Gurevich commonly known as the MiG. I further dedicate this book to tho participants in the projects described herein whose names remain unknown due to the secrecy and compartmentalization that existed at the time.

Acknowledgements

I acknowledge all of you with whom I served at Area 51, those named and those whose names I cannot reveal for national security reasons. In either case, thank you for your service. Your participation saved the lives of countless aircrews then, now, and into the future.

TABLE OF CONTENTS

Declassification:	8
PREFACE	**9**
The need to know your enemy	9
About the book, MiGs Over Nevada	9
Why the need for aerial assets?	12
The Cold War - A new type of war.	12
The role of the Central Intelligence Agency	13
The CIA base needed the new Soviet MiG to exploit	13
Secretly testing a foe's aircraft	16
Why Area 51	18
The evolution of the Groom Lake Facility	22
Why the CIA and not the Air Force or Navy operated Area 51	27
Area 51 - A business	33
The Air Force Foreign Technology Division.	34
CHAPTER 1 - Loss of American air superiority	**38**
Putting guns back on the US fighter planes	40
A new type of aerial warfare was not the only problem	40
Was it the planes?	41
What did the exploitation team at Area 51 have?	45
Test environment and procedures	45
What the exploitation team had to work with	49
What the exploitation team added for the technical and tactical evaluations	51
CHAPTER 2 - The CIA Area 51 Groom Lake operating facility	**52**
Area 51 customers	52
CIA commanders of Area 51	52
Area 51 Special Projects Exploitation team cadre	53
Radar cross section evaluation of the MiG-21	62
CHAPTER 3 - HAVE DOUGHNUT Tactical Evaluations	**64**
HAVE DOUGHNUT exploitation team leaders	64
The pilot Familiarization Flight	64
Tactical evaluation of the MiG-21	67
Joint Air Force and Navy technical evaluation flights on the MiG-21	74
AIM guided missiles	76
Radar homing and warning set AN/APR-25	76
The MiG-21 30 mm cannon	76
The MiG-21 Armament and Fire Control System	78
TAC inventory vs. MiG-21F FISHBED flight test overview	79
Emphasis on what was learned from matching the F-4 against the MiG-21	80
USN MiG-21 findings	81
Results and Exploitation team Discussion MiG-21	84
Maintenance discrepancies	88
Summary of technical evaluation of MiG-21 unique design features	88
Summary of tactical operational applications	88
The exploitation results recommended:	88

AFFTC lessons learned	88
ADC comments on Project HAVE DOUGHNUT	88
Findings: the bottom line	89
What the planes should do if encountering a MiG-21	90

CHAPTER 4- MiG-17F Projects HAVE DRILL/HAVE FERRY — **91**

Background of the MiG-17F technical exploitation	93
Discrepancies found during technical exploitation.	95
Description of the MiG-17 test item	96
The MiG-17F pilots	96
The shock from exploring the MiG-17	96
Technical exploitation of the MiG-17F FRESCO	97
Technical exploitation objectives	99

CHAPTER 5- Performance, Stability, and Systems Evaluation — **113**

Foreign Technology Division cockpit evaluation	115
MiG-17F profile	116
MiG-17 Combat flights at Area 51	119
Radar Cross Section evaluations	127
AFFTC and USN MiG-17F propulsion systems evaluation	129
Operational test exploitation of the MiG-17F	131
Naval Air Propulsion Test Center VK-1F engine static transient ground test	133
Preflight phase (HAVE FERRY)	133
Flight phase (HAVE FERRY)	134
Manhours expended for maintenance on the HAVE FERRY MiG-17	134
Scope of the HAVE FERRY evaluation	134

CHAPTER 6- Tactical evaluation of the MiG-17F at Area 51 — **135**

The HAVE DRILL and HAVE FERRY maneuvers	135
F-4 PHANTOM II vs. MiG-17F FRESCO C	135
F-100 vs. MiG-17F FRESCO C	140
F-5A vs. MiG-17F FRESCO C	144
F-105 vs. MiG-17-F FRESCO C	146

CHAPTER 8 - MiG-17F FRESCO exploitation conclusions — **154**

Navy recommendations	154
Tactical flight summaries	155
Maintainability (HAVE DRILL)	157
MiG-17 Deficiencies noted	157
TAC summary of the test program of the MiG-17F FRESCO C	158

CHAPTER 9 - Other MiG Projects in Nevada — **162**

Air Force assuming CIA operations at Area 51	162
Project HAVE PRIVILEGE	163
Project HAVE PAD	163

Epilogue — **167**

Constant Peg and the Red Eagles	168
Red Hats	173
Project HAVE GLIB	174
Defections with MiG planes continued	174
MiGs in Nevada exposed	175

Other countries copying USAF and USN's lead in aerial combat training	175
United States Navy TOPGUN	178
United States Air Force Red Flag exercise	179
Exploitation at Area 51 developed Air Force and Navy training programs	179
References	184
The MiG Bandits of Nevada	185
US planes lost in Vietnam	188
About the Author	197
Other Books by Author:	199
GLOSSARY	**202**

Declassification:

The contents of this book are declassified. The National Air Intelligence Center declassified the formerly Top Secret HAVE DOUGHNUT, HAVE DRILL, and HAVE FERRY FTD reports referenced in this book in September 1997 after review by NAIC/PA and NAIC/TAA to comply with IAW Executive Order 12958. FTD, TAC, Navy, AFFTC, SAC, ADC, and ASD for DIA referenced as HAVE DOUGHNUT and HAVE DRILL/HAVE FERRY: FTD-CR-20-13-69-INT Vol. I, "HAVE DOUGHNUT Technical," 1 Nov 1968, FTD-CR-20-02-69 Vol. I, "HAVE DR1LL, HAVE FERRY Technical," 1 Nov 1969, and FTD-CR-20-02-69 Vol. H, "HAVE DRILL/HAVE FERRY Tactical," 1 Apr 1970.

On October 29, 2013, the CIA posted its role in the projects in a release identified as The Area 51 File: "Secret Aircraft and Soviet MiG Declassified Documents Describe Stealth Facility in Nevada National Security Archive Electronic Briefing Book #443."

http://www2.gwu.edu/-nsarchiv/NSAEBB/NSAEBB443/

The author presented this book to the CIA PBR for review March 10, 2014.

PREFACE

The need to know your enemy

At the end of the Korean War in the 1950s, USAF pilots of the North American F-86 Sabre alone had a kill ratio of 10:1 against the MiG planes. A decade later, the US found its kill ratio reversed in the Vietnam War even before encountering the latest Soviet-built MiG-21 Fishbed, the Top Gun of the Communist world, elevating the US loss ratio to 9:1. Believing the reason being the enemy having superior planes, the logical recourse was to acquire and exploit the enemy's plane to discover what made it superior. The US acquired the enemy's planes and to its dismay found most United States Air Force and 100% of the United States Navy pilots losing their first simulated air battle against the Soviet MiG-17 or MiG-21. This happened at Area 51. The reasons and answers are the focus of *MiGs Over Nevada.*

During the 2006 declassification of the Constant Peg program, the USAF held a series of press conferences about the former top-secret US MiG programs in Nevada. Declassification revealed that in addition to the classified exploitation flights at Groom Lake in 1968, the US flew the Soviet-built MiG more than 15,000 sorties, training approximately 7,000 aircrew against dissimilar MiG aggressors in the Nevada desert between 1980 and the end of the program in 1988.

An October 2013 Central Intelligence Agency declassified document release enables the publishing of this book *MiGs Over Nevada.* This book is about recently declassified projects at Area 51 that reinstated the air superiority of the United States Air Force and Navy following a dismal record of defeat in the Vietnam War. This recent declassification released this insider to publish an account about how the CIA with FTD (Foreign Technology Division) exploitation teams of specialists addressed this disastrous military weakness by exploiting the Soviet MiG-21 and MiG-17. The participants included AFSC (Air Force Systems Command), the Laboratories at Wright-Patterson AFB, the AFFTC (Air Force Flight Test Center), SAC (Strategic Air Command), NDA (National Defense Agency), NASIC (National Air and Space Intelligence Center), NATC (Naval Air Test Center), the NWC (Naval Weapons Center), and the TAC (Tactical Air Command) using CIA contractor specialists and their facilities at Groom Lake, Nevada.

About the book, MiGs Over Nevada

MiGs Over Nevada is an insider's account of the exploitation of the MiG-17F FRESCO C, MiG-21F FISHBED, and the MiG-23 FLOGGER and how the projects inspired decades of exploitation projects that morphed into operations known as the Red Eagles and Red Hats. The results of the projects also inspired the establishing of the Navy's Top Gun Fighter Weapons School, the USAF Weapons School, and the USAF's Red Flag training exercises.[1]

---

[1] Red Flag and TOPGUN are quite different, even if the reasons for creating them were similar. TOPGUN is the Navy Fighter Weapons School located at Fallon NAS in Nevada. Originally a select few fighter pilots with an air to air mission went there to become weapons and tactics instructors in fighter squadrons. The Air Force has a similar program, initially called Fighter Weapons School, now just Weapons School. Both schools have since been

MiGs Over Nevada details the HAVE DOUGHNUT, HAVE DRILL, HAVE FERRY, and HAVE PAD technical and tactical exploitation projects to determine vulnerability, performance, stability, propulsion, instruments, radar cross-section, visibility, acoustics, infrared, and control of their warplanes against the MiG-17, MiG-21, and MiG-23. SAC evaluated the performance of the MiG-21F-13 against the ECM systems of the B-58 and B-52, Air Force and Navy engineers and scientists evaluated the aeronautical systems during assembly and disassembly of the MiG-21 and MiG-17 project vehicles, and AFFTC engineers evaluated performance information of various instruments.

However, due to a high national security classification, little was known until the release of documents in October 2013 about CIA contractor EG&G, the central participant. EG&G provided its Area 51 Special Projects cadre of specialists to conduct the tracking, data retrieval and storage, using its control rooms, ground based VHF, C, S, and X-Band radar systems, and state-of-the-art data collection and processing equipment. The author was a member of this exploitation team.

MiG-21 FISHBED flying over Nevada

---

either merged with other schools (Navy) or expanded (AF) to cover other missions. Each school is academically challenging and lasts about four months. Red Flag is quite different. It's a large-force employment exercise that lasts about two weeks, held several times a year at Nellis. Air Force, Nevada Navy, Marine, and allied aircrews all play at Red Flag, so that lots of aircrew (and a few GCI controllers) can get experience in the crowded skies of a major conventional war. In contrast, the weapons schools train a relatively small number of crews, and don't let people from other services or other nations play.

MiG-17F FRESCO flying over Nevada

MiG-23 flying over Nevada

Why the need for aerial assets?

Any military leader will say that an army's best weapon is knowledge of its enemy's forces. Throughout the history of man, reconnaissance required visually observing and reporting on enemy presence and activities.

Expressing the need for aerial reconnaissance, the Duke of Wellington, Napoleon's adversary at Waterloo, reportedly said, "The most difficult part of warfare was seeing what was on the other side of the next hill." In today's world, the most difficult part of warfare also includes seeing what is in the air.

In 1904, Orville Wright piloted an aircraft across Ohio's Huffman Prairie Flying Field, covering a distance of 356 feet. Between 1904 and 1916, a series of aviation firsts took place at Huffman Prairie, i.e., first controlled turn-first circle-first controlled bank and,-first figure-eight, making it the world's first permanent flight school. This prompted General Werner von Fritsch, Commander-in-Chief of the German Army in World War I to predict, "The nation with the best aerial reconnaissance facilities will win the next war."

True to this prediction, combat operations and strategic targeting during both World War I and World War II quickly embraced the airplane.

The Cold War - A new type of war.

The United States witnessed the capabilities of the atomic bombs on Hiroshima and Nagasaki in 1945 to "end" World War II, and awoke to find the USSR had turned adversary and Cold War contestant in a race for the "ultimate weapon of mass destruction."

In the late 40s and early 50s, NATO gained international prominence and the world realized radical change as Europe recovered from World War II. Germany split into East and West Germany; the Berlin blockade and air bridge began and ended; the Communists gained control of Czechoslovakia; and in 1949, the Soviets tested their own atomic bomb following in 1955 with tests of a hydrogen bomb. Both of these Soviet achievements occurred far sooner than projected by experts in Washington, thus the threat of nuclear weapons fed the mutual suspicions and reactionary foreign policies of both the United States and the Soviet Union.

The USSR aggressor, the Red army occupied the Easter European countries and refused to withdraw even decades after World War II had ended. The USSR was not satisfied merely with occupying the countries that the Red Army had overrun during the war. It wanted to take over Greece and Turkey through communist insurgencies in these countries. It took a show of military force by the US to persuade Stalin to pull the Red Army out of parts of the country of Iran under occupation. Later, when the Soviet Union sent in more troops to suppress rebellions in Hungary, East Germany, Poland, and Czechoslovakia, the whole world came to realize that the governments in these countries were truly nothing other than puppet regimes of the Soviet Union, which enjoyed no moral legitimacy. Where the United States was determined to carry democracy to the rest of the world, the Soviets were determined to spread communism.

Rumors of the Soviet Union having vast bomber forces shocked Washington into seeking information on Soviet strategic forces and their real strategic capabilities. Paranoia about the great Evil Empire, the Russian Bear spread through the American populous, causing ordinary American citizens to imagine their seeing flying saucers and to believe the Russians joining up with some extraterrestrial invaders. Fear of the intentions of the USSR after World War II caused American imaginations and paranoia to lose control. American citizens constructed nuclear bomb shelters in their back yards, expecting any day to see parachutes floating earthward carrying armed troops waving the hammer and sickle flag of the Soviet Union.

The role of the Central Intelligence Agency

Sharing this paranoia and the memories of Pearl Harbor, President Truman ordered the United States Air Force separated from the Army Air Corps and signed the National Security Act of 1947 to establish the National Security Council. The National Security Agency, in turn, reactivated the old World War II OSS to become their modern day Central Intelligence Agency with a Director of Central Intelligence answering directly to the president. Scientific intelligence production in the CIA started with the Scientific Branch in the Office of Reports and Estimates only to merge the branch with the Nuclear Energy Group of the Office of Special Operations to form the Office of Scientific Intelligence (OSI) on December 31, 1948.

The Central Intelligence Agency initially collected and analyzed information gathered by other agencies, including the various branches of the military, but as the Cold War escalated, branched out to perform operational functions and intelligence gathering on its own. The 1949 passing of the CIA Act stipulated classification of not only the CIA's activities, but its budget as well. Additionally, any other government agency could transfer funds to the CIA without regard to any provisions of law. The Act further stipulated that if the CIA's activities were exposed, the US government could "plausibly deny" any responsibility for the CIA's actions.

In its early years, OSI faced opposition from outside the CIA with regard to the extent of its authority over other elements of the intelligence community and the technical analysis of foreign weapons systems. From its inception to the present, the CIA competed with the military service intelligence organizations, a split that begin when the operations elements of the CIA refused to take OSI direction with regard to the collection of scientific intelligence.

During the next three decades, Edwin Land, Polaroid's chief executive officer, served as a key adviser to the CIA's science and technology effort. Chairing the intelligence committee of President Dwight Eisenhower's Technological Capabilities Panel, Land urged a reluctant Allen Dulles, the Director of Central Intelligence, to pursue development of a special high-altitude aircraft to overfly the Soviet Union and obtain detailed photographs of Soviet installations. This was the genesis of CIA-Air Force U-2 programs AQUATONE, subsequently CHALICE and IDEALIST of Lockheed's Skunkworks Works.

Five years later, 1952, the US government established the National Security Agency (NSA) within the Department of Defense to manage a national signals intelligence effort that included the communications intelligence activities of the military services. In 1958, the NSA became responsible for supervising military service electronic intelligence (ELINT) efforts, which failed to satisfy the CIA's need for ELINT, particularly intercepts of telemetry from Soviet missiles undergoing testing and Soviet radar systems capable of detecting CIA reconnaissance aircraft.

Consequently, to compete with the armed services, the CIA became heavily involved in aircraft, ground stations, ships, and eventually satellite electronic intelligence collection. In 1954, Allen Dulles approved the CIA's first ELINT plan. In 1960, the author became a participant in this EI collection.

The CIA base needed the new Soviet MiG to exploit

Craig Thomas was about 30 years off with his USSR technology when he wrote the book about the fictional jet plane in the 1983 movie picture "Firefox" starring Clint Eastwood. Interestingly, the MiG design produced for the movie closely resembled the stealthy aircraft appearing many years later. However, that was Hollywood and not the way of conducting the business of acquiring enemy aircraft. In the real world, it was not that easy to acquire an enemy's prized weapons. If you could not capture it on the field of battle, you bought it.

One of the most amazing of the reward campaigns was the attempt to steal a combat-ready Russian MiG-15 Fighter for one hundred thousand dollars during the Korean War.

The Soviets had designed the new fighter just after WWII. It was a high-altitude interceptor able to reach almost Mach 1, maneuverable at high altitude, armed with cannons, and had the ability to stay in the air for over 1 hour. The Soviets powered it with their copy of the British Rolls-Royce jet engine that had a higher thrust than the original. Its performance was superior to that of any Western fighter. The MiG-15 totally outclassed the American P-51 Mustangs, F-80 Shooting Stars, and the F-84 Thunder jets. The Americans had to wait until December 1950 for the arrival of the swept-wing F-86 Sabre-jet. Even then the MiG-15 climbed faster, and was every bit as maneuverable.

The name of this mysterious plot was Operation Moolah, the Korean War effort to entice a Communist pilot to fly a MiG-15 fighter to an allied airfield for a reward of $100,000.

There are several versions of how this campaign came to be. Regardless of which scenario was correct, Operation Moolah (GI slang for money) was ready by April of 1953. On 1 April, the UN Joint Psychological Warfare committee approved Operation Moolah using data sheets of the Headquarters, 1st Radio Broadcasting & Leaflet Group, 8239 AU, APO 500 dated 20 April 1953. The United Nations Command offered $50,000 to any pilot who flew his MiG-15 to the south. It offered an additional $50,000 to the first pilot who took advantage of this offer.

The campaign used both radio broadcasts and distribution of aerial leaflets in the Russian, Chinese, and Korean languages. The US believed at the time, and later proved that all three countries provided pilots for the air war over North Korea. Each of the three countries (The USSR, China, and North Korea) operated their own air force units as if they were fighting against their enemies. No pilot served in another country's air force unit. All the MiG-15s, however, were marked with North Korean insignia (A red star inside red and blue circles).

In 1953, North Korean pilot, Lieutenant No Kum-Sok, defected with a new Soviet built MiG-15, joining a returning flight of F-86s and landing unnoticed at Kimpo Air Base. In doing so, he received $100,000, the bounty offered in Project Moolah to any pilot delivering a MiG to the United States. After No surrendered his aircraft, the USAF took the plane to Okinawa, where the USAF gave if USAF markings and test-flew it by Capt. H. E. "Tom" Collins and Maj. Chuck Yeager. The Air Force then shipped the MiG to Wright-Patterson Air Force Base after an unsuccessful attempt to return it to North Korea. It is currently on display at the National Museum of the United States Air Force.

In 1954, Lieutenant No immigrated to the United States where he anglicized his name to "Kenneth H. Rowe." The US evacuated his mother from North Korea to join him in the US. He subsequently graduated from the University of Delaware with degrees in mechanical and electrical engineering. He married an émigré from Kaesong, North Korea, raised two sons and a daughter, and became a US citizen, working as an aeronautical engineer for Grumman, Boeing, Pan Am, General Dynamics, General Motors, General Electric, Lockheed, DuPont, and Westinghouse.

In 1970, No learned from a fellow defector that, as punishment for his defection, North Korea executed his best friend, several family relatives, his battalion commander, vice battalion commander, the battalion's political officer, the air division's chief weapons officer sponsoring No's Communist Party membership, his regimental commander, and the commander of the North Korean First Air Division.

Lieutenant No's defection was a dream come true for US intelligence who considered this the most important technical intelligence of the 1950s because it validated US methods.

Renowned test pilot Chuck Yeager test flew this aircraft and began a trend that was to last nearly forty years. Within three years, no less than four Polish pilots flew their MiG-15s to freedom, enabling the USAF to test fly the MiG-15 at Wright Patterson, Eglin, and PAX River against the B-47, B-36, and F-86.

The success of the MiG 15 exploitation sparked a continuing quest for possession of enemy planes. In virtually hundreds of cases warring nations have offered cash to the enemy — sometimes for defections or weapons, other times using cast for aid to friendly personnel or to purchase loyalty to a friendly government. During the Vietnam War, the US offered gold to North Vietnamese soldiers and Viet Cong Guerrillas to persuade the enemy to aid American pilots forced down over Communist held territory.

Early in the Vietnam War, US Aircraft dropped leaflets signed by the Commander, US 2nd Air Division, offering $35,000 in Vietnamese currency for the safe return of Lieutenant George E. Flynn shot down while flying an A1E Skyraider over the Dong Thai outpost, Hieu Le District, Kien Giang Province.

Secretly testing a foe's aircraft

There was nothing new about testing a potential foe's aircraft in secret. In the early 1950s, the USAF tested a Yakovlev Yak-23 at Wright-Patterson Air Force Base, Ohio, before packing the fighter into a C-124 Globemaster and returning it to a co-operative owner in Eastern Europe. The tests of this Yak-23 remained secret for more than 40 years after they took place.

The Soviet VVS had an opportunity to study a pair of US F-14 Tomcats and extensively test them against current Soviet fighters in the middle of a desert just outside of Moscow-at the Zhukovsky Flight Test Center, known then as "Ramenskoye" airbase because of the nearby Ramenskoye highway. The Tomcats could have only come from Iran.

Another Russian organization, the Central Aerohydrodynamic Institute, known by its Russian abbreviation "TsAGI" consistently participated in testing of foreign aircraft and their components. Among the aircraft tested at this facility were the two F-86 Sabres captured by the VVS in Korea and numerous other foreign aircraft obtained during various local conflicts, in particular during the Korean and Vietnam wars, as well as reconnaissance aircraft shot down over or near the USSR. The Moscow Aviation Institute or MAI frequently participated in this research.

Soviet-built Mikoyan-Gurevich MiG-21, the most advanced Soviet fighter plane at that time, production began in 1959, during which time Egypt, Syria and Iraq received numerous planes. The Israeli Mossad undertook Operation Diamond (Hebrew: Mivtza Yahalom) to acquire a MiG-21. The operation began in mid-1963 and ended on August 16, 1966, when an Iraqi Air Force MiG-21, flown by a defector, landed at an air base in Israel.

In 1964 an Iraqi born Jew, Yusuf, contacted Israeli services in Tehran (Iran and Israel had diplomatic relations at the time) and Western Europe. Since he was 10 years old, Yusuf had worked as a servant for a Maronite Christian family. His girlfriend's friend was married to an Iraqi pilot named Munir Redfa. Redfa was annoyed that his Christian roots prevented his promotion in the military. He was also upset about orders to attack Iraqi Kurds. Yusuf understood that Redfa was ready to leave Iraq.

A female Mossad agent befriended Redfa. He told her that the Iraqis forced him to live far away from his family in Baghdad; his commanders did not trust him, and allowed him to fly only with small fuel tanks because of his Christianity.

Redfa traveled to Europe to meet with Israeli agents who offered him $1 million, Israeli citizenship, and full-time employment. Israel sent numerous Mossad agents to Iraq to assist the transfer of Redfa's wife Betty, their two children aged three and five, his parents, and a number of other family members out of the country. Betty and their two children went to Paris for what she thought was a summer vacation. They took the other family members to the Iranian border, where Kurdish guerrillas helped them to cross into Iran, and then to Israel.

The opportunity to defect came about on August 16, 1966. Jordan radar detected Redfa flying over northern Jordan where Syria assured them that the plane belonged to the Syrian air force and was on a training mission. Two Israeli Air Force Dassault Mirage IIIs met Redfa's plane as it entered Israel, and escorted him to a landing at Hatzor. Later at a press conference, Redfa said that he had landed the plane on "the last drop of fuel".

Soon after his defection, Redfa's MiG was renumbered 007, reflecting the manner in which it had arrived. The Israelis analyzed the jet's strengths and weaknesses against IAF fighters. In May 1967, director of CIA Richard Helms noted that Israel had proven that it had made good use of the aircraft, when on April 7, 1967, during aerial battles over the Golan Heights, the Israeli Air Force brought down 6 Syrian MiG-21s without losing any of its Dassault Mirage IIIs.

In January 1968, Israel loaned the MiG to the United States, which evaluated the jet under the HAVE DOUGHNUT program. The transfer helped pave the way for the Israeli acquisition of the F-4 Phantom, which the Americans had been reluctant to sell to Israel.

The HAVE DOUGHNUT MiG-21 w/Israeli markings before delivery to Area 51

Why Area 51

A key factor in establishing the venue for flight-testing the U-2 spyplane was the security and activity offered by the atomic testing program in Nevada. On December 18, 1950, President Truman approved the use of a section of the Nevada Bombing and Gunnery Range for the Atomic Energy Commission to begin testing small yield nuclear weapons on continent. While the first test detonated on January 21, 1951, the February 12, 1952 signing of Public Land Order 805 officially established the Nevada Proving Ground. This original site consisted of 435,000 acres to include Frenchman Flats, most of Frenchman Lake (Dry), Yucca Flats, Yucca Lake (Dry) and Rainier Mesa. Eventually, the two national laboratories received specific sections of the site for their test programs and area designations put in place.

The name Area 51 conveniently adopted the numbering system of the AEC along with the creation of Areas 1-17 at that time. Area designations provided a means of calculating subsistence payments to the work force. For example, in the early 1960s, Reynolds Electrical Engineering Co., Inc. (REECo) routinely paid its personnel one wage if they worked in Mercury,

Nevada (the base camp) and more if they worked in the forward areas. Las Vegas based personnel were not paid subsistence. REECo time cards from this period contained a two digit location code (01 for the test site) and a two digit area code for the area that an employee worked in each week.[2] In the mid 1950s, AEC and NASA acquired another portion of the Bombing and gunnery range for use in the nuclear rocket program designated the Nuclear Rocket Development Site (NRDS). Later Public Land Orders led to the expansion of the list and workers at the AEC's TTR facilities used Area 52 on their time cards. Again, the secrecy related to TTR activities have captured a map designation, in this case Area 52 to play off the Area 51 name.

In 1955, the Central Intelligence Agency (CIA) needed a test facility for the Lockheed U-2 spy plane codenamed Project AQUATONE. For security reasons, CIA officials did not believe that they should fly the new U-2 spy plane at Edwards Air Force Base, California and sought a more secure venue. Richard M. Bissell, Jr., CIA's director of the U-2 AQUATONE program, with his Air Force liaison, Col. Osmond J. "Ozzie" Ritland, reviewed dozens of potential test sites to test fly the U-2 spy plane. They selected an airstrip, called Nellis Auxiliary Field No.1, located just off the eastern side of Groom Dry Lake, about 100 miles north of Las Vegas. It was also just outside the Atomic Energy Commission's (AEC) nuclear proving ground at Yucca Flat.

In early 1955, the U.S. Air Force made Nellis Auxiliary Field No. One on the Las Vegas Bombing and Gunnery Range available for Lockheed's use in testing their new U-2 aircraft. Bissell secured a presidential action adding the Groom Lake area to the AEC proving ground. This, in effect, constituted Area 51's birth announcement.

The CIA, Lockheed's chief customer, designated the location Watertown Strip. An agreement between the U.S. Air Force and U.S. Atomic Energy Commission authorized the AEC's M&O contractor, to construct a runway, dorms, and a few other buildings at the location. The CIA crafted a press release pertaining to the work that it released in May of 1955. A follow-on AEC press release identified the location as the Watertown strip. Maps of the Nevada Proving Ground during this period show "Watertown" beyond the Northeast corner of what was to become the Nevada Test Site. In mid June 1957, the CIA placed Watertown in an inactive status.

---

[2] The Area 51 security badges were identical to the AEC badges except they contained the number 8, which provided access to the Groom Lake facility.

Watertown Lives

References to Area 51 had largely displaced references to Watertown by the 1980s. Watertown was still in existence, however. A September 1994 study conducted for DOE entitled "Development of the Town Data Base: Estimates of Exposure Rates and Times of Fallout Arrival Near the Nevada Test Site" included Watertown as a populated location in Lincoln County.[1] The study was conducted to determine levels of contamination that could have occurred as a result of nuclear detonations on the Nevada Test Site (now Nevada National Security Site).

Page 53 of the report lists the various locations within the Alamo sector of Lincoln County which, in the interest of completeness, includes Watertown, as well as the Groom Mine.

As of 1994, Watertown still existed in a formal DOE report.

County	Township	Location Name	UTM	Lat (°N)	Long (°W)	Ctrl #
	Union	Winnemucca	11TMR3935	40.961	117.730	53
Lander	Austin	Austin	11SMP9471	39.485	117.075	772
		Laboard Ranch	11SMP9756	39.350	117.040	835
Lincoln	Alamo	Alamo	11SPM6336	37.353	115.165	767
		Ash Springs	11SPM6047	37.453	115.196	770
		Buckhorn Ranch	11SPM6527	37.272	115.144	788
		Crystal Springs	11SPM5655	37.526	115.240	846
		Dodge Constr. Camp	11SPM1465	37.622	115.713	808
		Groom Mine	11SPM1033	37.334	115.763	823
		Hiko	11SPM5762	37.589	115.227	825
		Lincoln Mine	11SPM2166	37.630	115.634	843
		Southpaw Mine	11SPM4370	37.663	115.384	892
		Tempiute	11SPM2068	37.648	115.645	2010
		Watertown	11SPM0928	37.289	115.775	908
		Whipple Ranch	11SPM5866	37.624	115.215	909

---

[1] Development of the Town Data Base: Estimates of the Exposure Rates and Times of Fall Arrival Near the Nevada Test Site, Carol B. Thompson and Richard D. McArthur, Desert Research Institute, September 1994, page 53.

Named after CIA director Allen Dulles' birthplace in Watertown, New York, the CIA established and constructed Watertown, a fledgling base with a single, paved 5,000-foot runway, three hangars, a control tower, and rudimentary accommodations for test personnel. The CIA included a few amenities that included a movie theatre and volleyball court along with necessities such as a mess hall, several water wells, and fuel storage tanks. CIA, Air Force, and Lockheed personnel began arriving in July 1955 and Richard Newton of the CIA assigned as base commander 1955-1956, followed by Landon McConnell 1956-1957.

Nuclear weapons testing at nearby Yucca Flat directly affected test and training activities at Watertown by its close proximity, requiring all workers to wear radiation badges to measure their exposure to fallout, and to evacuate the base prior to each detonation.

Testing of the U-2 plane wound down and CIA pilot classes of three CIA detachments and an Air Force pilot class completed their training. Watertown became a virtual ghost town serving as a laboratory to determine the nuclear shielding qualities of typical building materials found in any American town. Nearby, the AEC detonated the 37-kiloton PRISCILLA shot at Frenchman Flat followed by the WILSON shot. A truly spectacular atomic shot, HOOD, the sixth nuclear shot of Plumbbob, substantially damaged the Groom Lake airbase with a shockwave shattering windows on two buildings and buckling t metal roll-up doors on the hangars.

Even as other CIA and DOD projects moved in and the facility's name changed with the projects, the Atomic Energy Commissions, later the Department of Energy, continued to refer to the Groom Lake facility as Watertown.

Public Land Order 1662, dated June 20, 1958, added the 38,400 acres of land, to include Groom Lake, to the Nevada Test Site. Seventeen months later, the AEC announced the construction of new facilities for "Project 51." the AEC's M&O support contractor performed the work and provided follow-on support at the site. Because of the requirement for a two-digit area code on the REECo employee timecards, Project 51 morphed into "Area 51" for timecard purposes and that term began being used to identify the softball team and on maps and in other documentation. As evidence of the term's common acceptance, NTS phone books listed Area 51 phone numbers and local television stations lists of their transponders included the ones at Mercury, Area 12, and Area 51.

Dramatic changes came to Area 51 two years later with the advent of CIA Project OXCART to develop the Lockheed A-12 proposed successor to the U-2. The A-12 Mach 3, high-flying reconnaissance plane was America's first stealth plane, which required the radar cross section (RCS) test capability operated by a special projects exploitation team established for the CIA at Groom Lake by Edgerton, Germeshausen & Grier (EG&G), the prime contractor for the AEC atomic bomb testing. Codenamed, Project 51, the Central Intelligence Agency expanded the original Watertown runway and facilities in preparation for the Oxcart Program. The CIA produced a plan for construction and engineering using the CIA cover story of an engineering firm preparing the USAF support facilities to conduct radar studies. An EG&G subcontractor, Reynolds Electrical and Engineering Company (REECo) upgraded the Watertown facility under Project 51, bringing in housing, and adding support facilities.

Base construction began on 1 September 1960 and continued on a double-shift schedule until 1 June 1964. C-54 aircraft ferried in workers from Burbank and Las Vegas. The existing 5,000-foot runway was incapable of supporting the weight of the A-12, so the CIA constructed a new airstrip (runway 14/32) between 7 September and 15 November 1960, completing essential facilities by August 1961.

The CIA obtained, dismantled, and erected on the base's north side three surplus US Navy hangars, designated as Hangar 4, 5, and 6, built a fourth, Hangar 7, and converted the original U-2 hangars to maintenance and machine shops. The agency transported some 400 US Navy surplus Babbitt housing units to the base and made them ready for occupancy of personnel living at the facility Monday to Friday as well as those required to be on base during weekends and holidays.

The facilities in the main containment area included workshops and buildings for storage and administration, a small commissary, control tower, and fire station. On 15 November 1961, CIA's Werner Weiss, a.k.a. the Desert Fox and his wife Velma arrived in Las Vegas, Nevada for Werner to become CIA's third commander at the base. The first flight of the CIA A-12 spyplane occurred in April 1962.

Construction of the Area 51 facility reached completion in 1965, expanding the site population to 1,835 personnel with contractors working three shifts a day. Lockheed-owned Constellation and C-47 aircraft made several flights a day ferrying personnel from Burbank and Las Vegas to Groom Lake. Hughes, Honeywell, Pratt & Whitney, and Perkin-Elmer set up shop on the base to support the CIA Oxcart project. CIA provided its own security and communications, contracting EG&G to operate the mainstay of the facility, the Radar Target Scatter [RATSCAT] one-of-a-kind facility combining the best in monostatic and bistatic radar cross section (RCS) measurements similar to one located at White Sands Missile Range in New Mexico.

With the arrival of the HAVE DOUGHNUT MiG in 1968, Area 51 gained a new nickname, DREAMLAND, purportedly derived from DREAMLAND, a poem by Edgar Allan Poe. It describes lakes that "endlessly outspread" with waters "lone and dead." More to the point, Poe admonishes that "the traveler, traveling through it may not-dare not openly view it; Never its mysteries were exposed, to the weak human eye unclosed." Coincidence or not, it was certainly an apt description of Area 51. The CIA replaced Werner Weiss with Richard A. "Dick" Sampson in 1968 at the end of Project Oxcart and the beginning of the MiG projects that are the subject of this book. Following Sampson, Sam Mitchell commanded the base until April 1979 when it transferred to the control of the United States Air Force Detachment 3, AFFTC at Edwards commanded by Lt Col Larry D. McClain.

The evolution of the Groom Lake Facility

Over-all Shot Of Watertown Area From Hills Shooting Northeast.

1956

1968

EG&G Special Projects building and radar array for tracking and RCS

1968

Groom Lake in 2007

Why the CIA and not the Air Force or Navy operated Area 51

OXC-3146
Copy 2 of 7

14 MAR 1962

MEMORANDUM FOR : Deputy Director (Research)

SUBJECT : Discussion of Air Force Versus Agency Operation of Project OXCART

REFERENCE : Document Dated 14 October 1960; Subject: "Organization and Delineation of Responsibilities--Project OXCART" (OXC-0321)

1. Discussion of the relative merits of locating OXCART operational control in the Air Force as opposed to CIA is at once both extremely difficult and disturbingly simple. The difficulties are associated with sorting out from nearly seven years of experience in Project IDEALIST the most salient features forming the background of that essentially successful joint Air Force-Agency operation. The disturbing simplicity comes from the same Project IDEALIST experience which confirms the belief that CIA did not apply any magic formula in devising a workable arrangement with the Air Force in the conduct of that Project. The whole picture is further complicated by the fact that there is in existence at the present time a basic document referred to above, which was signed nearly two years ago by Mr. Dulles as DCI and the then Chief of Air Staff, General Thomas D. White, agreeing to retention of operational control for Project OXCART by this Agency in accordance with the salutary experience gained in Project IDEALIST. The critical paragraphs of the referenced document read as follows:

"3. b. The Project Headquarters will be responsible for any continued research and development, operational planning, and the direction and control of activities in the final operational phase of the Project when overflights are being launched.

"4. c. The third phase will be that of active operations. This phase follows the decision as to operational readiness. The final decision as to execution and timing of actual overflight missions shall rest jointly with the CIA and the USAF, subject to such guidance as may be received from higher authority, and in accordance with notification, coordination, and support procedures employed in Project OILSTONE.* The line of command shall be direct between operational units and the CIA."

*Project OILSTONE is the Air Force code name for IDEALIST.

2. It is worthy of note that the signatures of General White and Mr. Dulles on the reference represent the culmination of careful discussions between elements of the Development Projects Division and the OXCART liaison office in USAF Headquarters, which in turn reflected the views of USAF operational elements up to and including the Chief of Air Staff. While we do not argue that the arrangements established by the reference cannot be changed merely because they have previously been agreed to, we do feel that it is worth considering that the conclusions of two years ago wherein operational control was vested in the Agency for OXCART deserve careful restudy at this time.

3. Basic to the consideration of where operational control should lie seems to us to be the status of United States policy with regard to the conduct of overflights of denied areas, particularly the USSR and the Soviet Bloc. To our knowledge there has been no formal indication that the United States Government is prepared to accept the risks inherent in having such overflights conducted by the military arm. Assuming for a moment then that it is safe to conclude that CIA sponsorship as opposed to that of the Air Force offers desirable political benefits to the present Administration, let us proceed to identify as well as we can those advantages and disadvantages inherent in continuation of the presently agreed arrangements which places operational responsibility in CIA.

4. **PERSONNEL SELECTION METHODS.** We should like to review briefly here the manner in which Air Force officers and enlisted men are acquired from USAF for duty with Project OXCART. In the first place the top command elements in the OXCART field detachment were all selected on a personal basis by CINCSAC. He has personally nominated from his own wide acquaintance in SAC eleven key officers who form the operations cadre in the detachment. These include the Detachment Commander, the Director and Deputy Director of Operations, the Director of Materiel, and the principal members of the Operations Staff, including four senior Operations Flight Planners and three Navigation Flight Planners. All of these officers are key men with many years of combat and reconnaissance experience in USAF, beginning with the Detachment Commander, who was the senior reconnaissance officer in SAC prior to being detailed to this Agency. For the balance of the Air Force complement of the Detachment and its supporting elements, Air Force officers on detail to DPD did the bulk of the detachment manpower planning for the OXCART Project. It was they who established the criteria from a professional standpoint in accordance with the highest Air Force standards. Nominations for the Project are solicited by DPD through the Agency's Military Personnel Division, which translates them into requisitions levied upon a Special Projects Personnel Section in Air Force, responsive to the OXCART liaison office in USAF Headquarters. The Special Projects Personnel Section directs Air Force commands to nominate individuals for a highly classified special assignment without defining its exact nature. Written instructions go to the commands, insisting that "only the best qualified candidates be nominated". Eventual nominees are reported to MMPD in CIA, which conducts its own

OXC-3146
Page 3

screening of applicants based upon experience with the type of operation conducted in DPD in the past. One out of three nominees in the airman ranks received by MMPD under this system is screened out before the list reaches this Staff. Additional screening is conducted by the DPD Personnel Section, in close consultation with experienced Air Force detailees in the Division. Of the applicants finally put in process, one out of four is proven less than acceptable, either as a result of thorough security checks, polygraph, or medical examinations. All military nominees complete psychiatric tests upon entry into DPD and Project OXCART, and psychiatric approval may be required if the written test warrants it in the opinion of the Medical Staff. The Agency, therefore, is applying its own rigid entrance requirements to the already select list of nominees made available to us by USAF. It is a generally fair statement to say that in the security and medical areas CIA adheres to stricter requirements than would be the case were Project OXCART to be staffed by Air Force in accordance with their criteria. It is worthy of noting, for example, that the officer promotion rate among Air Force detailees to DPD is considerably higher than average within the Service. In nearly five years, for example, only one officer among the many detailed to DPD has failed a promotion after entering the primary zone of consideration. In summary on this point then, the evidence suggests that officers and airmen currently concerned with Project OXCART, both in Headquarters and the field, are nominated, screened, and selected in accordance with the highest standards of the United States Air Force and CIA.

5. **UNITARY OPERATIONS CONTROL.** It seems to us that one of the most persuasive arguments for the retention of operations control in this Agency is that by doing so we will have ensured not only a desirable close working relationship between research and development on the one hand and actual flight operations on the other, but in addition we will have taken advantage of an extremely shortened command line, which in the language of the referenced document "shall be direct between operational units and the CIA". As we now envision it, this direct channel which avoids cumbersome DOD mechanisms would flow for mission approvals direct from the White House to the Director of Central Intelligence, thence to the Deputy Director (Research), and onward to the field unit through the Chief, Special Projects Division and the Control Center in the Division. Unless and until one has become enmeshed in the complexities of DOD policy and program review and approvals mechanisms, he cannot fully appreciate the tremendous advantages of direct and uncontaminated command communications. It is a truism to point out that this shortened channel offers one of the principal benefits to be derived from Agency operations control: maximum security of the mission. As a separate but closely related benefit, preservation of Agency control offers the White House the option of deniability to a degree not present should the enterprise be undertaken and operated under military aegis. Unitary operations control as conducted in the Agency offers another useful and meaningful dividend in the form of reduced manpower required to plan, support, and conduct the operational mission, both in Headquarters and in the field. Actual operations planning for the entire OXCART Program will be

accomplished by a total of about thirty officers, split almost evenly between Washington and the field detachment. This figure is ridiculously low when plotted against a comparable standard Air Force operation, and is made possible only through minimizing of unnecessary and bureaucratic procedures, shortened command lines, and timely Air Force support in areas ancillary to operations; i.e., weather forecasting, communications systems, and tanker support. The caliber of this assistance is not only due to personalized support from highest Air Force command levels but by the unusually high priority granted the Project by the USAF itself. The biggest bonus of all, we feel, is that unitary control offers tremendous advantages in promptly relating the operations people to those responsible for research and development at the same piece of real estate and under the same roof. Since the so-called R&D phase will overlap by months or possibly even years the inception of the operations phase, there will need to be a continuous interface between developers and users. In the Air Force, with its multitude of related but separate commands, operators would most certainly be obliged to report their conclusions and recommendations to developers only after various levels of review had been accomplished, thus increasing the number of time-consuming decisions required.

6. **UNITARY PROJECT MANAGEMENT.** Anything that has thus far been said about the desirable aspects of unitary operations control it seems to us apply with equal force to over-all project management, from inception to delivery of the end product.

Given the strong language of the President's charge to the Director in his 17 January letter, establishing Mr. McCone without a doubt as the central focal point in the United States Government for all intelligence activities, as well as his own expressed desire to retain mission planning responsibility for OXCART in the Agency, it seems to us that the present well-exercised machinery for blending COMOR requirements, Agency development and operations responsibility, and NPIC handling of the end product offers the greatest assurance that the Director will at all times be the first to know the most about the OXCART intelligence product.

Contrast, if you will, the situation existing today in the SAMOS Program with that of Project CORONA as being an analagous situation to that which might exist were this present COMOR/CIA/NPIC combination to be disturbed. As an Agency, we know little or nothing about target programming for SAMOS. We know only a little more about launch parameters, orbital characteristics, or recovery techniques. End product processing is also not scheduled to be done at NPIC but at Westover Air Force Base. In the event of a successful SAMOS shot, the DCI must rely upon the DOD to tell him what transpired, what was covered in response to which requirements, and he has no immediate and direct access to a properly interpreted end product as he does in CORONA and IDEALIST at the present time.

Unitary management of the OXCART reconnaissance capability, whereby the entire cycle of funding, contracting, research and development, flight testing, operations or collection and end product processing all come under one roof, appears even on the surface to be so attractive in support of the Director's given responsibilities as to be well nigh irresistible. Should the Air Force be inserted only in the operations phase, it seems obvious that any credit for a successful operation would in the first instance fall to them; in the event it should not be successful, Air Force could argue that it was beyond redemption by the time they moved in. We submit that at all levels USAF operational inputs are receiving primary consideration and that from our experience, they see the benefits to the entire intelligence effort which are the by-product of the system established in Project IDEALIST some years ago.

7. **SECURITY**. One very major benefit derived from single source control of OXCART is the tested Agency security system. It is obvious that any move to split responsibilities between the Agency and the Air Force at the operational threshhold will automatically necessitate more clearances of USAF personnel. Under the present system of tight control in a single security complex, Air Force directs support and assistance from its various commands with minimum spread of knowledge. Were they to become wholly responsible for operations, they can be expected to follow their normal in-house practice of more fully informing commands participating in the effort.

USAF respect for Agency security standards and practices is reflected in their wide acceptance of the ▓▓▓▓ system and their stated desire to enlarge it to encompass other sensitive programs. Our experience over the years has been that in Air Force operations management of their own programs in reconnaissance they do not exercise the same degree of security hold-down that we have traditionally employed.

In the field of physical security, both the Air Force and we agreed early in our joint relationship that reliance should be placed on CIA's assets. The Agency necessarily exacts higher professional standards in its security staff than does the Air Force Provost Marshal, plagued as the latter is by constant rotation and generally low skill levels among available airmen, few of whom voluntarily choose the field of physical security. Any decision to place the OXCART Area under exclusive Air Force control would necessitate reliance upon traditional USAF Provost Marshal forces with what we feel is a lessening of effective security control.

8. **SUMMARY**. Much has already been said about what we regard as the inevitable pressures that would result on development were operational control to revert to Air Force in OXCART. In our view this continues to be a concern,

OXC-3146
Page 6

and one which can only deteriorate with the passage of time, since development will not end with the onset of operations. It also seems highly unlikely that mission planning responsibility can be retained by the Agency while yielding operational control to the Air Force, since there is such a close interaction between the two. As we have said, under the presently approved definition of responsibilities between CIA and the Air Force set forth in the reference, those wholly desirable contributions which must be made by USAF if the program is to succeed are already provided for. It <u>will</u> be the best qualified Air Force operations officers who <u>will</u> be exercising operational control over OXCART. In our opinion the only tangible difference will be that they will be responsive in the first instance to Agency management and objectives, which after all could not long continue to be so different from those of the Air Force that there is much to be gained in complicating an already smoothly functioning relationship by splitting the program in its most critical phase. On these facts it seems wise to rest our case.

25X1A9a

Colonel, USAF
Acting Chief, DPD

Distribution:
1 - DD/R
2 - ASST CH/DPD
3 - AC/DPD
4 - C/DPD/DB
5 - DPD/SPB
6 - DPD/SO
7 - DPD/RI

Area 51 - A business

Located within Area 51 of the AEC's Atomic Proving Grounds known today as the National Nevada Security Site, the CIA base at Groom Lake came to be known simply as 'Area 51,' a mere name that today conjures up images of government conspiracy and unexplained mysteries, dating back to the Cold War between the United States and the Soviet Union. Over the years, this non-existent air base acquired several identities, most of them to conform with the activities of the current customer: Area 51, Groom Lake, Dreamland, Nevada Test Site, Nellis Test Range, Paradise Ranch, the Ranch, Watertown Strip, Red Square, and the Pig Farm. Its mailing address became Pittman Station, a former, one-room post office in a rundown area along Boulder Highway in Henderson, Nevada. The pilots flying the North Range of Nellis referred to Area 51 as The Box because of the box shape of the restricted area shown on their navigation maps.

Declassification of various projects have revealed the accomplishments of individual projects, but told little about how these accomplishments came about—why they occurred at Area 51 with little ever said about the projects merely being those of customers using the facilities at Area 51. Once the project concluded, the customer moved out and another customer moved in.

The National Classified Test Facility at Groom Lake owes its existence as an Air Force base largely to the foreign materiel exploitation (FME) programs. These began in the late 1960s while the CIA still operated Area 51 as joint CIA-DIA-USAF-USN FME projects. The USAF FTD at Wright-Patterson AFB, Ohio, was the lead agency customer of the CIA facility.

Few have ever wondered about the cadre at Area 51 who serviced the customer. The reason was simple. The CIA operated the base with the EG&G Special Projects exploitation team more highly classified than the atomic bomb Manhattan Project. The customer came to Area 51 because of this venue having the only technical means and knowledge to service the needs of the customer.[3] The Central Intelligence Agency had developed the most highly technical laboratory on earth for flight development and testing, and electronic combat development and testing. It was at this CIA facility that the USAF FTD made testing and evaluation of foreign aircraft technology the longest continuing United States classified military airplane program.

Air Force Systems Command had large divisions that handled different aspects of technology: Aeronautical Systems Division, Electronic Systems Division, Armament Division, and others developed new technology for the AF. Foreign Technology Division was (and still is, as NASIC) involved with analyzing foreign military hardware, particularly foreign aircraft, some air defense systems, and space systems. They figured out how foreign fighter jets worked based on any info they could get: signals from the fighters, radar tracks of the fighters, photographs, and maybe human sources, whatever they could get. Engineers getting their hands on real hardware enabled them to fill in gaps in knowledge and confirm things they pieced together.

In 1968, FTD was part of the Intelligence Community, as NASIC is today. Intelligence organizations had an intelligence community 'chain of command', separate from their military chain of command. The CIA director was Director of Central Intelligence, and the CIA was only one small part of the community under his (loose) control. Beneath him was a Director of Military Intelligence, who ran the Defense Intelligence Agency, and somewhat controlled all

---

[3] It is not generally known even among the Special Projects team at the time if they continued operating as consultants for the CIA or somewhere along the line reverted to being DOD.

intelligence activities in the DOD. There were service intelligence chiefs as well. The intelligence chain handled decisions about production of intelligence reports amidst lots of cooperation (and rivalry) among the military services and civilian agencies. Once the intelligence community was involved, there was an honest effort to get all the interested players involved.

During the Cold War, secret test flying of Mikoyan-and-Gurevich Design Bureau (MiG) and other Soviet aircraft was an ongoing mission dating back to the acquisition of the first Soviet-built Yakovlev Yak-23 in 1953. This effort most likely continues today. Unlike the other "black" airplane programs, such as the Have Blue, Lockheed U-2, or the Blackbird family of Mach-3 planes, Foreign Aircraft Technology operations remain classified. Despite the declassification of the HAVE DOUGHNUT, HAVE DRILL, HAVE FERRY, HAVE PAD, and the Constant Peg program, the evaluation of foreign aircraft likely continues.

The Air Force Foreign Technology Division.

The Air Force FTD led much of the exploitation of the MiG in Nevada. FTD evolved from the Air Technical Intelligence Center (ATIC) established on May 21, 1951, as a USAF field activity of the Assistant Chief of Staff for Intelligence under the direct command of the Air Materiel Control Department. ATIC analyzed engine parts and the tail section of the Korean War Mikoyan-Gurevich MiG-15 obtained during Project Moolah. The Defense Intelligence Agency (DIA, created on October 1, 1961, and USAF intelligence organizations/units reorganized where ATIC became the FTD assigned to Air Force AFSC to provide intelligence estimates to the National Security Council through the 1962 United States Intelligence Board (the CIA's Board of National Estimates). In October 1993 at the end of the Cold War, FTD became the National Air Intelligence Center as "a component of the Air Intelligence Agency," and by 2005 had a Signals Exploitation Division after being renamed the National Air and Space Intelligence Center on February 15, 2003.

SECRET

CENTRAL INTELLIGENCE AGENCY
Washington, D.C. 20505

APPROVED FOR RELEASE
DATE: MAY 2002

25 July 1969

MEMORANDUM FOR: The Director of Central Intelligence

SUBJECT: Acquisition of Soviet Technical Manuals for the MIG-21PFM Aircraft

1. I wish to inform you that the Clandestine Service has just acquired three Soviet technical manuals for the MIG-21PFM [FISHBED F] interceptor. These manuals are the first documentary data to be acquired on this most recently deployed version of the MIG-21. The manuals cover performance characteristics, tactics, and maintenance.

2. The MIG-21PFM is a further development and improvement of the MIG-21PL (modified FISHBED D), for which the Clandestine Service obtained manuals previously. The MIG-21PFM is now the main production model of this widely deployed fighter, which is operational or becoming operational in North Vietnam, North Korea, Cuba, Afghanistan, India, the UAR, and other countries. This version is now the most modern interceptor available to the air defense forces of the Warsaw Pact countries, excepting the USSR.

3. The most useful manual is expected to be the Combat Employment manual, which updates tactics used by the modernized MIG-21 and which probably is the primary combat guidance for North Vietnamese MIG-21 pilots.

4. These manuals contain confirmatory and new data on the RS-2US (ALKALI) air-to-air missile, a booster starter system to permit rocket-assist take-off, and the first documentary information available on a Soviet boundary layer control system (SPS) for an aircraft. The latter two systems (RATO and SPS), not previously incorporated together into an operationally deployed Soviet aircraft, permit the MIG-21PFM to operate from poor quality runways and improvised airfields.

SECRET

5. Analytical elements of the Intelligence Community are being notified via normal distribution channels of the acquisition of the above documents and are being provided pertinent extracts from the documents in [redacted] The FBIS is currently translating the documents and the translations will be disseminated in the CSDB series to all pertinent consumers.

Thomas H. Karamessines
Deputy Director for Plans

SECRET

S E C R E T

Distribution: The Director of Central Intelligence

The Assistant to the President for
 National Security Affairs

The Director of Intelligence and Research
 Department of State

The Director, Defense Intelligence Agency

The Director, National Security Agency

Director, National Indications Center

The Deputy Director of Central Intelligence

Deputy Director for Science and Technology

Deputy Director for Intelligence

Director for National Estimates

Director for Current Intelligence

Director for Strategic Research

Director for Economic Research

Director for Scientific Intelligence

S E C R E T

CHAPTER 1 - Loss of American air superiority

At the end of the Korean War in the 1950s, USAF pilots of the North American F-86 Sabre alone had a kill ratio of 10:1. This changed in the next war as the US pilots faced enemy still shooting cannons at US planes armed with only air-to-air missiles. Yet, US forces could not consistently track low flying MiG on radar, and were hampered by restrictive rules of engagement (ROE) which required pilots to visually acquire their targets, nullifying much of the advantage of radar guided missiles, which often proved unreliable when used in combat.

The new generation of United States fighters had brought with it the development of air-to-air missiles, fighter aircraft, such as the US Navy's F4H Phantom II. Later redesignated the F-4 in 1962, the aircraft was the first fighter designed from the start with only air-to-air missiles, carrying both radar-guided AIM-7 Sparrow III and the shorter-range AIM-9 Sidewinder infrared-guided missiles. With the new missiles came the new attitude that dog fighting was obsolete. The air-to-air training given to new Navy and Marine Corps F-4 crews was extremely limited, involving about ten flights and providing little useful information. By 1964, few remained in the Navy and Marine Corps to carry on the tradition of classic dogfighting. Then came the Vietnam War.

The early years of the air war over North Vietnam showed the faith placed in missiles was terribly in error. Between 1965 and the bombing halt in 1968, the USAF had a 2.15:1 kill ratio. The US Navy was doing slightly better with a 2.75:1 rate. We lost an American F-4 Phantom II, F-105 Thunderchief, or F-8 Crusader for roughly every two North Vietnamese MiG-17 FRESCOs or MiG-21 FISHBEDs shot down. Worse than losing the 10-1 kill ratio enjoyed by the US aircrews of the Korean War was the continuously growing percentage of United States fighters being lost in air-to-air combat. The 1966 3 percent loss of US aircraft losses due to MiG rose to 8 percent in 1967, and then climbed to 22 percent for the first three months of 1968.

The emphasis on air-to-air missile interception left the fighter combat crews with only the sketchiest knowledge of dogfighting. Originally conceived as a naval fleet air defense aircraft, and later adapted as an Air Force fighter-bomber, the design of the F-4 made it ill suited for a tight-turning dogfight. In contrast to the lighter MiG-17, the F-4 was large and heavy. When making a tight turn, the F-4 lost energy and airspeed whereas the MiG-17's superior turning capability then allowed it to close to gun range. All too often, hits from the MiG-17's "outmoded" cannons would then destroy the F-4.

The Soviet sponsors and North Vietnamese air force commanders knew the MiG-21 FISHBED's limitations and planned around them. They never committed their fighters unless there was a good chance of success and subsequent escape. During that stage of the Vietnam War, 80 percent of the North Vietnamese air force kills occurred with the victims being unaware that they were under attack.

Much of their success was their using "Red Baron" tactics, initiating attack from a cross-course intercept. MiG-21 fighters vectored by ground control intercept radar from Chinese airspace to position behind the F-4 Phantoms bombing targets north of Hanoi. Known as "blow-throughs," the MiG launched ATOLL missiles and zoom back to political sanctuary in China when the F-4s pulled up from their target.

Following the success of the small, highly maneuverable F-86 day fighter in the Korean War, US fighter design had changed to emphasize maximum speed, altitude, and radar capability at the expense of maneuverability, pilot vision, and other attributes needed for close combat.

This trend reached its extremity in the McDonnell Douglas F-4 Phantom, which was the principal fighter for both the US Air Force and Navy during the latter part of the Vietnam War.

The F-4, though originally an interceptor for defense of the fleet against air attack — never executed such a flight because no US fleet has ever came under air attack since the beginning of the jet age. The F-4 interceptor design met that required of the fleet defense mission using rapid climb to high altitude, high supersonic speed, and radar-guided missiles to shoot down threat aircraft at long distance.

Combat Losses in SEA

USAF fixed-wing		USN fixed-wing		United States Army	
A-1 Skyraider	150	A-1 Skyraider	48	USA fixed-wing	
A-7D Corsair II-	4	A-3 Skywarrior —	2	OV-1A Mohawk	3
AC-119 Shadow/Stinger	2	A-4 Skyhawk	195	OV-1B Mohawk –	2
B-52 Stratofortress	17	A-6 Intruder	51	O-1 Bird Dog	297
B-57 Canberra	38	A-7 Corsair	55	OV-/C/D series Laos/NViet.	67
C-7 Caribou	9	EA-1 Skyraider	1	Bell 205 (Air America)	1
C-123 Provider	21	F-4 Phantom	75	AH-1G	270
C-130 Hercules	34	F-8 Crusader	157	BELL	1
F-4 Phantom II	382	RA-5 Vigilante	18	CH-21C	14
F-100 Super Sabre	198	RF-8 Crusader	19	CH-34	2
F-102 Delta Dagger	7	S-2 Tracker	3	CH-37B	1
F-104 Starfighter	9	**USN fixed-wing shore-based**		CH-37C	1
F-105D Thunderchief	283	C-47 Skytrain -	1	CH-47A	83
F-105F/G Thunderchief:	37	P-2 Neptune	4	CH-47B	20
F-111A "Aardvark"	6	P-3 Orion	2	CH-47C	29
HU-16 Albatross	2	**USN rotary-wing**		CH-54A	9
O-1 Bird Dog	122	SH-3 Sea King-	8	H-13D	3
QU-22 Pave Eagle	7			H-37A	2
RF-4C Phantom II-	76	**United States Marine Corps**		OH-13S	147
RF-101 Voodoo	33	USMC fixed-wing		OH-23G	93
U-3B Blue Canoe	1	A-4 Skyhawk	81	OH-58A	45
CH/HH-3 Jolly Green Giant-	25	A-6 Intruder	25	OH-6A	842
HH-43B Pedro-	8	C-117 Skytrain	21	UH-1	60
CH/HH-53 Super Jolly	17	EA-6A Intruder	2	UH-1A	1
AH-1 Cobra	7	EF-10 Skynight	51	UH-1B	357
HUS-1	75	F-4 Phantom	72	UH-1C	365
UH-1E Huey	1969	F-8 Crusader	21	UH-1D	886
CH-37 Mojave	1	KC-130 Hercules	4	UH-1E	90
CH-46D Sea Knight	109	O-1 Bird Dog	7	UH-1F	18
CH-53 Sea Stallion	9	OV-10 Bronco	10	UH-1H	1313
		RF-4 Phantom	4	UH-34D	176
		RF-8 Crusader	1		
		TA-4 Skyhawk	10		
		TF-9 Cougar	1		

At the height of the Vietnam War, with the skies filled with technologically advanced American aircraft from both the US Navy and the USAF, the air battles lacked the past glories in the 1950s skies over Korea's MiG Alley. Their aircrews faced carefully trained North Vietnamese pilots and competent warriors such as top ace, Nguyen Van Coc; credited with seven aircraft and two Firebee unmanned aerial vehicles destroyed. His aircraft victories included two Air Force F-4s, one Navy F-4B, two "Wild Weasel" F-105s, one F-105D, and the only F-102A kill of the war.

The North Vietnamese Air Force's first jet air-to-air engagement with US aircraft was on April 3, 1965. The NVAF claimed the shooting down of two US Navy F-8 Crusader never confirmed by US sources, although they acknowledged having encountered MiGs. Consequently, April 3 became "North Vietnamese Air Force Day." On April 4, 1965, the VPAF (NVAF) scored the first confirmed victories acknowledged by both sides. It shocked the US fighter community when relatively slow, post-Korean era MiG-17F fighters shot down advanced F-105 Thunderchief fighters-bombers attacking the Thanh Hóa Bridge. The two downed F-105s carried their normal heavy bomb load, making them unable to react to their attackers.

Also in 1965, the USSR supplied NVAF with supersonic MiG-21s, which NVAF used for high speed GCI controlled hit and run intercepts against American air strike groups. The MiG-21 tactics became so effective, that by late 1966, the US mounted an operation to especially deal with the MiG-21 threat. Led by Colonel Robin Olds on 2 January 1967, Operation Bolo lured MiG-21s into the air, thinking they were intercepting an F-105 strike group, but instead found a sky full of missile armed F-4 Phantom IIs set for aerial combat. The result was a loss of almost half the inventory of MiG-21 interceptors, at a cost of no US losses. The VPAF (NVAF) stood down for additional training after this setback.

Using the F-4 in Vietnam as a fighter rather than as an interceptor severely miscast the plane at great expense to the Air Force and Navy aircrews. Against inferior North Vietnamese pilots flying small, highly maneuverable MiG-21s, the air-to-air kill ratio sometimes dropped as low as 2:1, where it had been 13:1 in Korea.

Under the HAVE DOUGHNUT and HAVE DRILL programs, the exploitation team used the first MiG flown in the United States to evaluate the aircraft in performance and technical capabilities, as well as in operational capability, pitting the types against US fighters.

Meanwhile, US Air Force and US Navy airmen flying contemporary advanced aircraft, combined with a legacy of successes from World War II and the Korean War, revamped aerial combat because of Project HAVE DOUGHNUT at Area 51. With what they learned at Groom Lake, the Navy put designs to the drawing board for an entire generation of aircraft, with engineering for optimized daylight air-to-air combat (dog fighting) against both older, as well as for emerging MiG fighters.

Putting guns back on the US fighter planes

In 1968, Chief of Naval Operations (CNO) Admiral Thomas Hinman Moorer ordered Captain Frank Ault to research the failings of the US air-to-air missiles used in combat in the skies over North Vietnam. Operation Rolling Thunder, which lasted from 2 March 1965 to 1 November 1968, and ultimately saw almost 1,000 US aircraft losses in about one million sorties. Rolling Thunder became the Rorschach test for the Navy and Air Force, which drew nearly opposite conclusions. The USAF concluded that its air losses were primarily due to unobserved MiG attacks from the rear, and was therefore a technology problem. The service responded by upgrading its F-4 Phantom II fleet, installing an internal 20mm Vulcan cannon (replacing the gun pods carried under the aircraft's belly by Air Force Phantom units, such as the 366th Tactical Fighter Wing), developing improved airborne radar systems, and working to solve the targeting problems of the AIM-9 and AIM-7 air-to-air missiles.

A new type of aerial warfare was not the only problem

The first exploitation of a Soviet-built MiG-21F-13 (FISHBED E) fighter-interceptor occurred during a troubling time in US history.

Conflict between a conservative regime and a growing number of anti-Vietnam protesters had spread from the campuses to the streets, and federal agencies were secretly investigating the loyalty of American citizens.

It didn't help the US air war when the Johnson administration forbid targeting North Vietnamese airfields, parked aircraft, command centers, and main radar installations. The US entered the Vietnam War with a move away from cannon fire to air-to-air missiles only to find in the Vietnam War that the dogfight was alive and well and missiles were not yet ready to replace the gun. The Top Vietnamese ace of the Vietnam War claimed nine kills: seven manned aircraft and two UAVs. In just one day, in December 1966 the MiG-21 pilots of the 921st FR downed 14 F-105s without any losses. Another change was the much more extensive use of helicopters and spotter planes that provided easy targets for the enemy's MiG fighter planes.

The only aircraft to come close for the air superiority role in conventional warfare was the F-100, which came along too soon for the M-61, therefore was equipped with four M-39s. The F-104 and F-105 were the first two planes equipped with the M-61 Gatling gun. The Air Force brought the F-4 into the its inventory for the air superiority role without a gun, but because of the Vietnam War losses, the fighter employed the M-61 Gatling gun carried externally in the SUU-16 pod.

The AIM-7, designed as an anti-bomber weapon, did not have the broad range for firing or maneuverability needed in a fighter versus-fighter engagement. However, the AIM-9 kill rate was somewhat better, about twenty-percent, during the latter part of the 1965-1968 campaign. The Air Force F-4E with an internal gun did not make its debut in the war until 1968 and after what the Air Force and Navy learned at Area 51 during Project HAVE DOUGHNUT.

The Vietnam War was the first conflict that saw wide scale tactical deployment of helicopters. The enemy shot down over 8,000 helicopters, costing the US over 5,000 US helicopter pilots in Vietnam. The US lost 2,200 Bell UH-1 Iroquois (Huey) helicopter pilots in the war, not counting all the Cobras, CH-53, Chinooks, H21 Shawnees, Siouxs, Choctaws, Sea Stallions, Jolly other choppers used in the war such as the OH-58 Kiowas, OH-6 Cayuse, AH-1 Green Giants, CH46s, etc. This number does not include other crewmen aboard those shot down helicopters, men that were not pilots, just members of the chopper's crew, door gunners, crew chiefs, etc.

Not counting helicopter pilots and air crewmen, the US lost over 6,000 fixed-wing/propeller/jet US pilots and air crewmen, killed or missing during the Vietnam War. The USAF lost about 2,584 men and the USN lost about 2,555 men. The USAF and USN together lost well over 2,000 fixed wing aircraft in addition to the US Army losing the 8,000 rotor-wing aircraft.

Was it the planes?

Once again, the US needed to acquire a MiG to find out. The MiG-21F-13 mystery plane was no longer a mystery once it first flew over Area 51 in 1968. Note the USAF markings on the plane redesignated as an YF-110B

The HAVE DOUGHNUT story begins on 16 August 1966 in the Middle East a year after Operation Rolling Thunder, the sustained air bombardment of North Vietnam began. Monir Radfa, an Iraqi Air Force captain, took off in his Mikoyan MiG-21F-13 from Rashid AB, outside of the Iraqi capital, for what was supposed to be a local navigation exercise. Instead, he made a dash to the southwest at low level, intending to defect. The Jordanian RJAF's Hawker Hunters, being too slow at low level, failed to intercept him as he streaked low across their country. He entered into Israel, lowered his landing gear, and wagged his wings to two intercepting Israeli Mirage III fighters to signal his intentions. The Israelis escorted him to Hatzor AB, where they gave him asylum.

The single-seat, supersonic, single-engine, delta wing, swept-back tail MiG-21 FISHBED fighter jet was one of the most potent fighters in the Arab air forces threatening Israel. The Israelis immediately set about flight-testing the mid-wing monoplane for over 100 hours over the next 12 months, learning its strengths and weaknesses, and teaching the Mirage III pilots how to defeat the MiG in a dogfight.

Initially hesitant to share its prize with the United States, Israel eventually concluded an agreement brokered by the US Defense Intelligence Agency (DIA) to loan the MiG-21 to the US for study in exchange for the US allowing them to buy the F-4 Phantom II, the American front-line fighter of the day.

At the time, the Israelis had made several overtures to the Johnson Administration to purchase the Phantom only to have President Johnson rebuff them out of a fear of escalating matters in the Middle East. Having the MiG-21 gave the Israelis enough advantage that the Phantoms were on their way and the US was finally able to study up close its vaunted adversary in the skies of Vietnam.

The Mystery Jet – the MiG-21F-13 Fishbed

Project HAVE DOUGHNUT exploitation of a Soviet-built MiG-21F-13 (FISHBED E) fighter-interceptor occurred from 23 January to 8 April 1968. Projects HAVE DRILL and HAVE

FERRY exploited two Soviet-built MiG-17F FRESCO Q fighter-interceptors from 27 January to 30 June 1969. These recently declassified projects were the genesis of the Navy's Top Gun Weapons School, Air Force's Red Flag exercises, the recently declassified USAF Red Eagles Constant Peg program, the Red Hats (still classified), and other similar MiG exploitation programs, the details of which remain classified.

The public release of HAVE DOUGHNUT, HAVE DRILL, and HAVE FERRY information is limited strictly to the referenced reports in this book and to protect international agreements and national security specifically excludes any subsequent exploitation efforts. To the extent authorized, the author discusses declassified exploitation programs that continued another 20 years as follow on programs generated by the successes of HAVE DOUGHNUT, HAVE DRILL, and HAVE FERRY.

The highly classified tests were the responsibility of the FTD of AFSC, a predecessor of today's Air Force Materiel Command. The word `Have' in project names such as HAVE DOUGHNUT (MiG-21) Or HAVE BLUE (the prototype for the F-117 stealth fighter), etc., identifies the project as belonging to FTD, now known as AFS.

The Air Force disassembled and transported the HAVE DOUGHNUT MiG by a Lockheed C-5 Galaxy to the USAF's secret testing base at Groom Lake, Nevada. It was unloaded in Hangar 5 where assembly began the next day. USAF's FTD that was part of the AFSC based at Wright-Patterson AFB in Ohio inspected, adjusted, repaired, and operationally checked all systems.

AFSC assigned all of its programs with the code word prefix "HAVE." For example, the original stealth demonstrator aircraft that gave rise to the Lockheed F-117 Nighthawk was code-named "HAVE BLUE." The MiG-21 on loan from Israel was code named "HAVE DOUGHNUT."

The program called for two categories of flight-testing of the MiG. The first type concerned technical analysis-performance, flight envelope, engineering, structures, and so on. For this, the Air Force made preparations at Wright Patterson Air Force Base by developing a "black box" and acquiring instruments to record test results.

The CIA at Area 51 also prepared for the testing by ordering an X-band Nike radar system from Fort Bliss, Texas. At the same time, the author, Nike-trained Thornton D. "TD" Barnes, an Area 51 Project OXCART participant, transferred from the NASA X-15 High Range in Nevada to the Flight Dynamics Laboratory at Wright Patterson to develop and test instrumentation for the impending MiG exploitation at Area 51.

From Wright Patterson, where his cover was his performing integrity tests on the Apollo 1 space capsule, Barnes returned to Nevada for assignment with the Special Projects exploitation team at Area 51. He arrived at about the same time that the CIA negotiated with the US Army at Fort Bliss, Texas for one of the Nike Hercules radar systems in which Barnes had attended years of schooling while serving in the Army. The CIA acquisition of this Nike radar system occurred long after the closure of Project OXCART at Area 51, thus identifying it as direct participation in the MiG exploitation projects.

The planned exploitation of the MiG-21 involved exploitation during the technical phase, and choreographed tactical exploitation phase flying the MiG in mock dogfights and comparison flights against US fighter aircraft. Because AFSC/FTD's emphasis was technical in nature, most of the HAVE DOUGHNUT flying concerned technical analysis.

During the HAVE DOUGHNUT tactical evaluations of the MiG-21, the Air Force TAC, in joint participation with the US Navy and other government agencies, conducted an analysis of the MiG-21F-13 FISHBED E day fighter weapons systems. The joint effort determined its tactical capabilities as a weapons system against their respective warplanes in air-to-air environments that include evaluating the performance of the MiG-21F-13 against the ECM systems of the B-58 and B-52.

The USAF FTD conducted acoustic measurements, Photometric coverage, engine flight test data, and along with the Naval Air Propulsion Test Center and AFFTC, exploited the propulsion system. The Naval Air Propulsion Test Center conducted static transient ground tests, and Naval Missile Center evaluated the SIRENA Tail Warning Receiver, RSIU-3M VHF transceiver, and SRO-2 IFF equipment. Results of the technical exploitations included flight test and performance data, maintenance summary, system and subsystem characteristics, design features, and technological information.

The USAF Tactical Fighter Weapons Center conducted the tactical evaluation under the overall management of the FTD, evaluating both the MiG-21 and the MiG-17 as total weapons systems and operationally in a tactical environment to compare them with a variety of TAC aircraft that the Air Force expected to confront as a threat.

The US Navy likewise conducted a tactical evaluation to determine the capabilities and limitations of the FRESCO C in the air combat maneuvering environment against Navy combat airplanes, to define area of comparative strength and weakness, and to determine the tactics necessary for Navy combat airplanes to defeat the Soviet MiG-17 FRESCO C in the air combat maneuvering environment.

Both the Air Force and the Navy recognized the MiG-21 as a day fighter at medium and low altitude using intercept tactics where the MiG was GCI vectored into the rear hemisphere where they attacked in high-speed, single-pass attacks. Due to tactics of US aircraft deploying in SEA, Southeast Asia, the AF and Navy seldom encountered the MiG-21 in the high altitude, point intercept role. The exploitation at Groom Lake provided both the opportunity to evaluate the MiG planes in that environment as well in an air-to-ground environment.

Sanitized - Approved For Release : CIA-RDP33-02415A000500140001-9

25X1A2g
■-9224-68
Copy 6 of 8
21 AUG 1968

MEMORANDUM FOR: Chief, Applied Physics Division, Office of Research and Development

ATTENTION: ■■■■■■■■■■■ 25X1A9a

SUBJECT: Nike Hercules Radar

REFERENCE:
25X1A2g
a. Memo for Record from AD/M/OSA; dated 28 May 1968; Subject: Arrangements for Shipment of Nike Hercules Radar ■-9045-68)

b. Memo for Record from D/M/OSA; dated 17 June 1968; Subject: Shipment of Nike Hercules Radar" ■-9106-68) 25X1A2g

1. Reference B, Paragraph 4, states that ORD had budgeted for spare parts for the Nike Hercules radar. The M & S van, part of the radar system, contained a year's supply of spares for field support costing $53,650.72. The M & S van, including the spares, was assumed to be part and parcel of the radar system transferred at no cost to the Agency.

2. 25X9A5 Upon the movement of the radar system, Fort Bliss (Capt. ■■■■■■■■■■■■■■■■■■■■) insisted that the spares in the M & S van were to be transferred at cost to the Agency. This was later confirmed by Lt. Col. F. D. Burnett, DSC/LOG, Department of Army.

3. 25X9A5 The spare parts from the M & S van were returned to Capt. ■■■■■■■■■■■■■■■■■■ on 7 August 1968 and an authentic receipt document received and is filed with D/M/OSA.

25X1A9a

Colonel USAF
Deputy for Materiel, OSA

25X1A2g
Handle via ■■■
Control System

Sanitized - Approved For Release : CIA-RDP33-02415A000500140001-9

What did the exploitation team at Area 51 have?

HAVE DOUGHNUT was an export MiG-21F-13 (Article 74) with an aircraft manufacture date in the last quarter, 1963. The aircraft had approximately 135 hours on it—the engine had 165 hours. No ATOLL missiles were included in the deal, so the exploitation team substituted almost identical AIM-9B Sidewinders. An R-37F axial flow turbojet with 12,650 pounds max afterburner thrust powered the clipped delta wing planform with swept tail surfaces. The MiG-21 FISHBED E weighed 11,017 pounds empty, 17,286 pound at takeoff, and a maximum of 18,072 pounds. It had a wingspan of 23.47 feet, length (without pitot boom) of 44.2 feet, and a height of 13.5 feet. Its armament included one 30 mm cannon with 60 round capacity, two ATOLL missiles, and a total bomb load on all three stations of 3,300 pounds. Its maximum performance speed was 2.05 Mach with a 57,500-foot service ceiling, and a strike radius of 370 nautical miles with external fuel. It carried a fuel load of 4,600 pounds internal and 880 pounds in the centerline tank.

Test environment and procedures

The HAVE DOUGHNUT participants planned the technical exploitation of the HAVE DOUGHNUT MiG in advance at Wright Patterson, determining what to seek and how to obtain it using what one might describe as equipment that included a specially designed black box to extract the needed information.

Once the plane arrived at hangar 5 at Groom Lake, the exploitation team-installed the instrumentation needed for the exploitation. This included an oscillograph, 12 channels-nav light switch/cannon switch, Gyros-Pitch, Roll, Yaw plus rates–vertical tail, fuel Flow Meters-total and normal, Photo Panel-Airspeed, Altitude, Mach, Free Air Temp, & (in nose)-Clock, instrument panel-A-13 clock, airspeed, altimeter, Mach,-accelerometer, stop watch, engine fuel temp, Cockpit-two Triad 16 mm cameras, voice recording system, battery, and a UHF radio.

Having physical possession of the MiG-21 for the first time, two exploitation teams, one from the Naval Weapons Laboratory, and one from a contractor for NWC (Naval Weapons Center) made a detailed structural study of the project aircraft. The Falcon Research and Development Company, under contract to NWC performed the examination of the plane, finding it in keeping with recognized shoddy Soviet design philosophies, when it came to construction. However, other systems indicated a healthy state of Soviet technology.

A serious, yet almost humorous incident early in the program validated the warranted concern for the unexpected. We hooked the plane to our laboratory instrumentation, where on the day of the first engine run-up of the MiG-21, virtually everyone related to the project came to witness the event. Someone noticing everyone wearing their security badges with attached dosimeters and film badges suggested gathering up all the badges to prevent accidentally sucking one of them through the engine during the run up. The exploitation team gathered up all of the badges and gave them to one of the military spectators to hold during the test. With the MiG engine screaming at military power and the plane's brakes straining to hold the plane in place, everyone gravitated to the front of the plane—including the one holding all the badges. Instead of sucking a badge through the engine, the engine sucked all the badges through, causing severe damage to the impeller blades of a relatively new engine (165 hours) that had yet to fly at Area 51. Fortunately, some of the Pratt & Whitney J-58 engine engineers were still on site from Project OXCART and were able to repair the engine enough to fly the HAVE DOUGHNUT tactical flights.

Calibrated test instruments installed for exploitation

Airspeed Dial Face

Mach Meter Dial Face

Rate of

MiG-21-F-13 Fishbed cockpit instrument

Bellmouth installation attached to aircraft during engine airflow evaluation.

 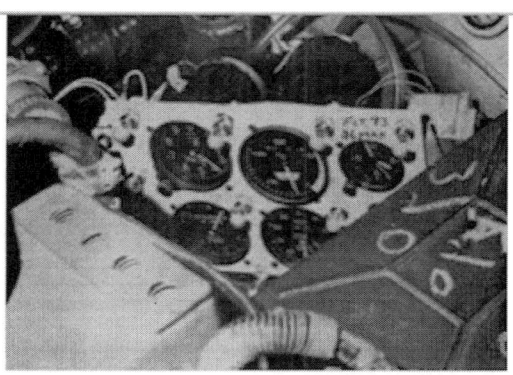

ADF Radio & IFF TransponderPhoto

Recording Camera Left SideRecording Camera

Photo RecorderAbrams B-9A

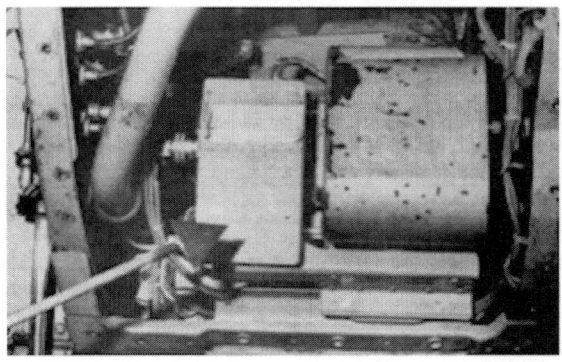 

Century Model 409 12-ChannelOscillograph

Voice Recorder

What the exploitation team had to work with

The MiG-21 fuselage was 4l.1 feet long (without pitot boom) with a wingspan of 23-5 feet, and weighed 11,100 pounds empty. For supersonic flight, it used a three position, translating, double angle inlet cone that extended from the retracted position to an intermediate position at 1.5 IMN (Indicated Mach Number) and to the full forward position at 1.9 IMN. The pilot could manually position the normally automatic cone from the retracted position to the 1.5 or the 1.9 IMN position.

Even with comparable technology levels, the applications varied appreciably. The Soviet approach placed emphasis on doing no more than necessary. For example, in the corners of the intake ducts where the flow velocity was low, the Russians did not countersink the rivets, leaving them protruding as much as one-eighth of an inch above the surfaces. This, however, caused negligible effect on the flow in the duct, and made significant savings in construction effort to the Soviets. Another method of simplicity and ease of maintenance for stowing the drag parachute cable channeled it through a shallow groove on the exterior surface of the aircraft along the ventral fin, holding it in place by simple harp-shaped spring clips with thin safety wires across their tops. The Soviet design concept 'wrapped' the smallest airplane possible around the available power plant to assure maximum speed, altitude, and acceleration performance. The Soviets carried this philosophy to the point where fuselage bulges at various places provided clearance for equipment and accessories, rather than increase the fuselage diameter or cross-section area. The performance evaluation of the MiG-21 attested to the success of this approach.

The MiG-21 power plant was a Type 37P twin spool turbojet equipped with a variable thrust afterburner. The engine, 181 inches long with a diameter of 35.7 inches, developed approximately 8,450 pounds of thrust at military power and approximately 12,650 pounds of thrust at maximum afterburner power.

The MiG-21 flight control system consisted of a manual rudder, hydraulically boosted manual ailerons, and an irreversible horizontal stabilizer. The main hydraulic system operated at 1200-1400 psi and powered the aileron boost and horizontal stabiliser. The plane did not provide a manual back-up control to the horizontal stabilizer, so to prevent over controlling at high speeds, an automatic control altered the gear ratio from the control level to the stabilizer to decrease the range of deflection required of the stabilizer. It provided no stability augmentation, aileron trim, or rudder trim.

The MiG-21's hydraulically actuated speed brakes were located on the underside of the fuselage, two forward and one aft. The landing gear was conventional tricycle type with selectable two or three wheel, air-operated brakes, using a lever on the control stick to activate braking and a rudder to control the desired wheel.

The aircraft was equipped with a three wheel, pneumatic, braking system, the added braking provided by the nose wheel brake making it very effective and further indicating the Soviet desire to operate their tactical fighters from relatively short runways.

Construction of the fuselage was rather unique, focusing on construction expediency as well as weight savings. Rather than building up with frames and an inner and outer skin or honeycomb material, the forward section was composed of a single piece of metal.

US standards for tactical aircraft, the range and payload capabilities of the MiG-21F-13 were very low, but the Soviets did achieve their apparent goal of developing a rugged, simple, highly reliable, and easily maintained fighter with exceptional climb, altitude acceleration, and maneuverability capabilities, surpassing any other aircraft operational in early 1960. These characteristics were still very good by tactical fighter standards if range and payload were not a prime concern.

The fuselage cells were non self-sealing bladder type with wing fuel stored in a "wet wing" with a fuel system capacity of approximately 4,500 pounds of JP-5.

The airplane used a simple, single button, self-contained, electrical battery starting system.

The airstart system incorporated an autonomous oxygen supply designed for restarts up to 39,000 feet. Sufficient oxygen was available for four to five airstarts of 30 seconds duration each. A special tank supplied aviation fuel during the starting cycle.

The MiG-21 cockpit was armor plated, including armor plating installed behind the pilot's seat, forward of the instrument panel and aft of the forward windscreen.

The sight system in the MiG-21 was an ASP-5ND lead computing gunsight. High Fix radar supplied range information in the cannon or rocket mode of operation to the gyro pipper. In event of radar failure, fixed range inputs were available from 650 to 6600 feet.

For fixed armament, the MiG-21 had one NR-30, 30-mm cannon faired into the fuselage under the right-hand side of the pilot's cockpit. The cannon had a linear action with a mechanical feed chute, which roughly followed the contour of the airplane's outer fuselage skin between the skin and the internal fuel tank. Firing rate was 850 rounds per minute with a muzzle velocity of approximately 2.500 feet per second.

The plane provided external stores on two wing-mounted stations and one centerline station. Each removable wing station carried one ATOLL missile, one bomb up to 1,100 pounds, or one 16-shot FFAR (Folding Fin Aircraft Rocket) pod. The permanently mounted centerline station carried one 130-gallon jettisonable fuel tank or a 1,100-pound bomb with no tail-warning receiver (SIRENA) installed.

Aircraft Modifications for Tactical Evaluation-Data: Modifications, to the MiG-21 were only those required for valid data acquisition.

What the exploitation team added for the technical and tactical evaluations

The exploitation team installed UHF communications and removed the standard VHF equipment while adding a UHF blade antenna,

To achieve representative combat, the exploitation team fabricated two wooden wing pylons and attached a LAU-7A missile launcher to each. The exploitation team then attached one AIM-9B training missile to each launcher.

The exploitation team placed a voice tape recorder on the right rear cockpit console and a communications lead that connected to the pilot's headset and microphone for providing necessary inputs,

Other modifications made for quantitative flight test data included a 10 channel oscillograph, two over-the-shoulder cameras, a photo panel, X band beacon, stopwatch, standard ITS instruments, airspeed indicator, calibrated Machmeter, and altimeter.

The ATOLL missile did not come with the plane, so for missile configuration, the exploitation team installed non-firing AIM-9Bs using an AIM-9 rail with laminated plywood/fiberglass and steel fittings. The substitution caused no performance change except slight improvement at low speed.

TECHNICAL ANALYSIS IN PROCESS ON THE HAVE DOUGHNUT. NOTE THE AIM-9B MISSILE INSTALLED TO SIMULATE

CHAPTER 2 - The CIA Area 51 Groom Lake operating facility

Area 51 customers

Until January 1968, the CIA Groom Lake facility had only one operational project and one customer — the Lockheed A-12 Project OXCART to develop a replacement for the CIA's U-2 spy plane. This changed in January 1968 as Project OXCART wound down and with the arrival of project HAVE DOUGHNUT, a joint USAF/Navy technical and tactical evaluation of the MiG-21F-13. The purpose and needs of this initial MiG exploitation project at Area 51 drew an entourage of interested and participating intelligence and technical agencies to follow the lead of the US Air Force and Navy. The services provided to these new customers by the CIA far exceeded the earlier RCS and flight tests during Project OXCART when OXCART was the only tenant of the facility.

Suddenly having multiple customers also changed the way the CIA conducted business at the facility. The customers arriving to work on the new MiG project did not have a need-to-know about the A-12 project and vice versa. Access to the different sections of the facility suddenly became restricted, as did social communication and networking on the base. The need-to-know and compartmentalization applied as well to the members of the Special Projects exploitation team of cadre where some worked for one customer and not the other. Housing suddenly segregated with each compartment establishing their own social and entertainment centers and activities. The residents suddenly muted any conversation related to work activities when they entered the shared mess facilities.

The customers for the MiG exploitation project hosted by the CIA at Area 51 included the USAF Foreign Technology Division (FTD) leading a exploitation team of specialists drawn from throughout the USAF and USN, including AFSC, the Laboratories at Wright-Patterson AFB, the AFFTC, SAC, NDA, NASIC, NATC, the NWC, and the TAC. Names of personnel involved remain secret to protect their privacy; however, declassification has authorized personnel involved in these specific projects to discuss data within the reports released by the various governmental agencies conducting the MiG projects.

CIA commanders of Area 51

- Richard A. "Dick" Newton (1955-1956)
- Landon McConnell (1956-1957)
- Werner Weiss (1958-1961)
- USAF Col. Robert J. Holbury (December 1961 to July 1966) with Weiss as deputy
- USAF Col. Hugh "Slippery" Slater (July 1966 to 1969) with Weiss as deputy
- Richard A. "Dick" Sampson (1969-1971)
- Sam Mitchell (1971-1977)
- Larry D. McClain (1977 to 1 April 1979) - first USAF site manager/detachment commander

Area 51 Special Projects Exploitation team cadre

Housed and transported separate from any other tenants, members of the Special Projects exploitation team reported only to their individual customer. Many of those serving and being unique to Project OXCART left Area 51 at the same time as their project. Named below are those CIA EG&G Special Projects cadre at Area 51 who remained for the MiG exploitation projects-and identified as exploitation team Organization #6300.

The addition of new highly classified projects tightened the compartmentalization and need-to-know environment within the Special Projects exploitation team, splitting the exploitation team up to support projects needing individual expertise rather than the exploitation team as a whole when OXCART was the only action going.

The fact that the exploitation team members did not talk about their work with each other did not mean that they did not get to know each other to the extent they played together as friends and families. Most knew in basic terms what the other did from giving them a hand with something. Nonetheless, they did not ask what, why, who, or how if they lacked the need to know. This did not mean that the team members lacked curiosity. It was quite the contrary; however, each member, with his own specialized ambitions and professional goals, expected protection from snooping regarding his own projects. The members assisted each other as needed, but totally respected the proprietary rights of each other's thoughts and achievements. These were not time-clock workers, but highly motivated individuals instead, each individually selected for his pristine ethics, his needed specialty, and his drive to achieve the impossible. To them, what they did at Area 51 was not work – it was a duty, privilege, and a pleasure.

Most being married with two children, and having common interests created the necessary bonding for shared national security concerns. This developed a cohesiveness where they worked together all week and then played together at the lake or on the mountain during the weekends. Rarely did they talk shop, and even under these conditions, they never did so with the wives or children present.

Speaking from my personal experiences, I, the author learned over time that each of the exploitation team and their families underwent the same security and compatibility evaluations, though many of them evolved from the atomic testing side. Consequently, almost every member the Special Project exploitation team carried both an AEC Q security clearance and a DOD top-secret security clearance.

Each member of the Special Projects exploitation team selected for his and her (there was only one "her"-Denise Haen based in the Las Vegas office) unique qualifications, family stability, ethical, integrity, and moral qualities as well.

For example, John Grace, a member of America's Who's-who, was responsible for obtaining many of the radar systems in the radar array. Denise Haen handled the administrative needs from recruiting, security, safety, human resources from the EG&G building on Sunset

Road in Las Vegas under Burt Barrett. Denise also held the distinction of being the first female to hold the title of Director of Special Projects Administration. Denise later retired from Special Projects after 31 years of service.

Dave Haen specialized in telemetry during Project OXCART. During Project HAVE DRILL, he became involved with a new radar system. As with most of the exploitation team, from that point on his career remains classified top secret. Over the course of over 30 years in Special Projects, Dave advanced to Director of Site Support Operations.

Jim Freedman came from the atmospheric testing of the atomic bomb at the various remote areas of the world to the Special Projects exploitation team as an administrator and courier. Declassification revealed that each evening Freedman dropped by CIA commander Werner Weiss's office to pick up the dispatch to Langley. He dropped it off to a United Airlines employee at McCarran International Airport in Las Vegas, and the next morning repeated the procedure in reverse —delivering the dispatch from Langley to Werner Weiss at Area 51.

NOT TO BE DISSEMINATED
OUTSIDE OSA-DD/R

SECRET

Approved For Release 2002/06/24 : CIA-RDP75B00326R000200120005-6

OKC-4560-63

27 FEB 1963

MEMORANDUM FOR THE RECORD

25X1A SUBJECT: [redacted]

1. On 12 October 1962, the undersigned, in the company of Messrs Cunningham, [redacted] and General Flickinger participated in the recruitment of Subject for Project OXCART. 25X1A

2. The first day followed pattern previously established and went pretty much according to script. Individual was initially interested in performance characteristics and, although his detailed questions were parried, he did obtain an admission that speed, on the conservative side, might be said to exceed MACH-2.

3. During the interview, one could see that General Flickinger was not impressed with Subject. This stems from concern over very inquisitive, intelligent, and domineering wife. Secondly, physiological findings, although disqualifying, which pertained to spatial bifida were a cause for concern as earlier raised by Colonel Ledford and General Flickinger.

4. On 16 October 1962, [redacted] and undersigned met Subject at the Shoreham Hotel for an answer. He indicated a great deal of concern about hiding the truth from his wife and, although it was first thought he was using this as an excuse, as the interview progressed, it was quite clear that this would be a source of emotional stress. This, together with the other considerations, placed [redacted] and the undersigned in the position of dissuading the individual from volunteering. At the conclusion, he was actually upset, but believed that his declination was made by himself and was in the interests of all parties concerned. 25X1A

5. Subject was given a security admonishment by [redacted] of not getting trapped into divulging what had transpired. Subject should forget that this recruitment effort had taken place. Termination security oath was executed by Subject. Subject could foresee no questions which he would not be able to handle. He will indicate that he had withdrawn from the voluntary program because it was not leading anyplace. Subject further indicated that there would be no need to explain why he had brought his flight gear on this trip. 25X1A

[redacted]
Chief, Personnel Branch

This document contains information referring to Project OXCART

Approved For Release 2002/06/24 : CIA-RDP75B00326R000200120005-6

SECRET

The termination of CIA Project OXCART in early 1968 reduced the number of Special Projects personnel to include personnel such as Harry Phiffer, who was in charge of the engineers at the site, Wayne Pendleton F-Systems (flight) manager went to work for Lockheed, and Jim Tarver G-Systems (ground) manager assigned elsewhere. Jules Kabat, assigned responsibility for the two radar systems in the antenna building: the S-band radar with the 60 feet dish and a Navy radar left the area to accept assignment in Vietnam.

Frank Harris, responsible for the C-band radar, and Bill McCloud, the Antenna Building lead tech, reported to Howard Schmit, the F-Systems senior technical supervisor, and George Percy, Senior technical supervisor for G-Systems.

Credit goes to the previous exploitation team for having in place this facility to test the MiG planes. Others included Dick Lampier-(G-systems engineer), Stan Busby-(G-systems engineer), Carl Newmiller-(draftsman), Bob Funknown-(Nike operator), Dick McEwen-(G-systems), Cowan Dawson-(C-band), Dick Wilson-(Q-bay), Robert Pezzini-(antenna building), Vern Williamson-(PPA), Dave Haen-(Q-bay), Sam Gamble-(draftsman), Jim Freedman-(admin), ? Helbert-(G-systems), Jim Cates-(G-systems), Eddie ?-(admin), Rocky ?-(G-systems Tech Supervisor), McGlothen-(Clothesline), CarCo C-47 pilots out of Albuquerque, New Mexico—Roy Kemp-Chief Pilot, Tom Hall-Joe Cotton, Hugh Starcher, Tom Losh, and Flo Deluna-aircraft mechanic.

The Central Intelligence Agency operated the Area 51 facility, however, few know outside the Special Projects exploitation team of the involvement of the Special Projects cadre at Area 51. Even within the Special Projects exploitation team, exploitation team members did not necessarily know what brought their contemporaries to Area 51. For illustration, the author will use his (Barnes') personal circumstance.

Sixty-five miles west of the Area 51 facility, Barnes was conducting the tracking for the CIA at Beatty because of his having the required security clearance from previous involvement with the CIA related to Project OXCART, the A-12 reconnaissance plane.

He attended and graduated over two years of schooling at the Nike Ajax and Nike Hercules surface-to-air radar and missile maintenance schools at Nike Ajax/Hercules: USARADSCHFTBliss, and the HAWK surface-to-air missile maintenance school at Hawk:USARADSchFBliss. This provided him with some unique radar experience needed at Area 51 for airborne high-speed RCS radar cross section evaluations of the America's first stealth plane, the A-12. He was experienced at tracking 3,000 mph missiles and one of only half a dozen experienced in tracking the hypersonic Mach 6.7 X-15 rocket plane using NASA's SCR-584 Mod-2 radar at Beatty, Nevada.

His job at the NASA radar site included other systems besides the radar, systems such as telemetry, DTS (data acquisition and transmission), analog to digital conversion, timing, and communications. NASA activities were overt and did not require a security clearance, thus his still having his security clearance from his Army service uniquely qualified him to serve the CIA need that required, on occasion, covertly using the NASA radar system for its ability to record a plane's velocity. Thus, he became a member of a classified element of CIA Project OXCART and the CIA/Air Force Project KEYLOCK referred to as the Seven Sisters. For these special flights, the NASA monitor at Beatty, Mr. Bill Houck, received a call requesting special track of an unidentified plane with the only recorded data being that of the station's velocity recorder. For these flights, he assumed the duty of radar operator and Houck provided the security to ensure other personnel at the station did not observe the track or data requiring special handling.

Six of the radar systems were Air Force SAGE ADC (Air Defense Command) and the radar at Beatty being a SCR-584 (short for Signal Corps Radio # 584) microwave radar developed by the MIT Radiation Laboratory during World War II. It replaced the earlier and much more complex SCR-268 as the US Army's primary anti-aircraft gun laying system.

It was for the combination of his unique qualifications and availability that the CIA embedded him into the Area 51 Special Projects exploitation team to conduct airborne RCS stealth evaluations of the A-12 and later on the MiG of Area 51.

Almost all engineers in the Special Projects exploitation team chose a senior technician classification to escape the "salaried" pay status of an engineer. As senior technicians, the exploitation team drew almost the same pay as engineers, except, after the first eight hours their pay rate increased to time and a half for four hours and then to double time straight through, 24 hours a day, until they arrived back in Las Vegas on Friday evening. Food and lodging were free and "someone" even furnished the exploitation team with snake proof boots for when they worked at the pylon out on the lakebed.

Area 51 honcho, Werner Weiss and his successor Dick Sampson of the CIA, accommodated the permanent party personnel at Area 51 with a small BX containing snacks and various personal hygiene items, swimming pool, exercise room, softball diamond, putting green, poolroom, and a three-stool bar called "Sam's Place," named after Dick Sampson.

The amenities reduced the hardship on the Special Projects exploitation team members maintaining two residents, one with their family in Las Vegas, and one at Area 51 where base housing assigned each a room in a row of Babbitt housing units. Each house contained a small living room where the exploitation team played poker and watched 8 mm movies played on a movie projector.

The Special Projects group usually banded in two groups for housing, one group being the boating enthusiasts with boats moored on Lake Mead[4] and the other being the Mt. Charleston[5] cabin dwellers with emphasis on the personnel selection including common interests. Further grouping often occurred to segregate the exploitation team according to project or customer. Extracurricular activities such as poker, rental movies, reading, and so forth, developed even more sub grouping according to interests.

CIA and other personnel staying at Area 51, such as Air Force, Lockheed, Hughes, Pratt and Whitney housed in similar houses, but clustered apart from the others. Very little association existed even in the Special Projects exploitation team outside one's professional group. For their off-duty pleasures, each customer developed their own entertainment, which involved converting a room in one of the residential houses into a bar and poker room. The OXCART group had what they called, "House Six," the Red Hat group, "The Red Hat." The permanent party personnel mainly used Sam's Place and the CIA amenities, except for poker games conducted in individual houses, which typically offered four bedrooms, a common living room, and kitchen.

Grouped even further was the author and three others who flew to and from Area 51 in a Queen Air plane, flying out of a secure compound along the side of the McCarran Airport along Sunset Road. Though unconfirmed, some believe the Central Intelligence especially chose the Queen Air group for their unique contributions to future projects rather than merely the current one.

A major amenity was the mess hall remaining open 24/7 and serving better food than any hotel or casino in Las Vegas. The Air Force support element during OXCART had arranged with the commissary officer at Wendover Air Base, Nevada to provide the mess hall with fresh oysters and lobster. Evidence of the great days of the CIA era remain today in the huge piles of oyster shells that the mess hall dumped on the Groom dry lake. Future archeologists will wonder how and when Area 51 was part of an ocean. Most famous was steak night where the base filled with authorized support personnel from Burbank and all over on Thursday for all you could eat

---

[4] Lake Mead - located on the Colorado River and the largest reservoir in the United States

[5] Mount Charleston, officially named Charleston Peak, 11,916 feet located about 35 miles northwest of Las Vegas

steaks.

Circa 1968, installation of a translator finally brought television to Area 51. Instead of providing a form of relaxing entertainment, it invoked outbursts of disgust from the engineers watching a great game of football, only to have the stronger signal of a bullfight out of Mexico override Channel 3 out of Las Vegas. The TV signal propagation changed with atmospheric conditions, causing skip conditions. Meanwhile, the poker games continued as the prime entertainment.

During Project OXCART, the Groom Lake evolved from a high desert surrounding a large dry lakebed surrounded by semi-barren mountain ranges within the Emigrant Valley to a fully functioning air base. Consequently, Project DOUGHNUT did not require installation or construction, leaving little for the permanent party personnel to do other than their specialty. Project OXCART had caught the spying eyes of the Soviet Union, the satellites that Area 51 personnel dubbed as ashcans. The arrival of the first MiG drastically increased the Soviet interest in the activities of Area 51, resulting in an increased frequency of satellites passing overhead. As with OXCART, the passing satellites required the ceasing of any electronic emissions and the hiding of any outdoor activities that included rushing any planes caught on the airstrip or tarmac into a hangar or hoot-n-scoot shed existing for that purpose.

Besides severely disrupting schedules and activities, the downtime because of passing satellites created a toll on the CIA's highly skilled and highly motivated personnel on the Special Projects exploitation team. To most, being unable to energize and signal emitting electronic equipment made their assigned project unbearably boring. This boredom, however, unintentionally developed new technology from the Special Projects personnel utilizing their knowledge, curiosity, and available means to experiment on something that later became a project at the facility.

Sadly, these achievements and most of the individual human accomplishments, mistakes (referred to as OS or "old shit!" moments, and humor have and may never become known. The author earlier mentioned the MiG engine sucking in all the film badges and dosimeters. Another occurred during OXCART when the Lockheed engineers raised the nose of the A-12 Blackbird plane, using a forklift, to evaluate fuel distribution. The fuel ran to the aft end of the plane, lifting the nose of the 105' plane into the rafters of the hangar. OS! A long-running incident attributable to the Special Exploitation team is one referred to merely as "the goat incident."

Following a rain, the Groom dry lakebed retained water for long periods. The lakebed did not absorb the water, so it stayed until it evaporated. The goat was an all-terrain, six-wheel drive, amphibious vehicle that the Special Projects exploitation team used to haul electronic test equipment and tools to the RCS pylon on the Groom lakebed. During a trip to the pylon, the vehicle somehow caught fire, becoming well advanced before the operator noticed it burning in the rear of the vehicle. The fire destroyed the goat and its contents. For years thereafter, any time a piece of test equipment became lost or misplaced; its loss was immediate attributed to the loss of the goat. Later, an inventory of all the equipment supposedly lost with the goat revealed losses that would have filled a small truck. (Most of items declared as lost usually showed up as being in use by someone else at the time)

The CIA A-12 Mach 3+ surveillance plane undergoing RCS evaluations on the pylon at Groom Lake. This was America's first stealth plane

Area 51 Special Projects Equipment

The author emphasizes that the Special Projects personnel, activities, and equipment discussed herein occurred half a century ago, and does not suggest that the same exists today; emphasis being that this book will discuss only what applies to the declassified projects. Having this equipment in place and operational at Area 51, plus the security it offered made it the obvious venue for the MiG exploitation projects.

The EG&G Special Projects exploitation team utilized six basic radar subsystems for cross section measurements of the MiG tactical phase of exploitation. An M-33, mobile, X-band fire-control radar, later replaced by the CIA's X-band Nike Hercules radar, tracked the target to generate range information and target bearings. A bull gear servo network manufactured in-house allowed the other radar systems to slave to the Nike's target bearing data

The Special Projects engineers and technicians operated and monitored the individual radar performances from a master control facility using only four of the radar systems in the measurement program. The VHF, C-band, G-Systems detected radar return hot spots, and F-Systems made radar cross-section measurements as it did on the previous project OXCART's A-12 (Blackbird) articles in flight at the threat frequencies. The S-band radar with the 60 feet dish DYCOMS (Dynamic Coherent Measurement System equipment gathered the reflectivity data.

Some of the Special Projects equipment utilized to control tactical missions, receive, process, and store data from the missions.

View of RATSCAT and EG&G complex 1969

Nike X-Band

 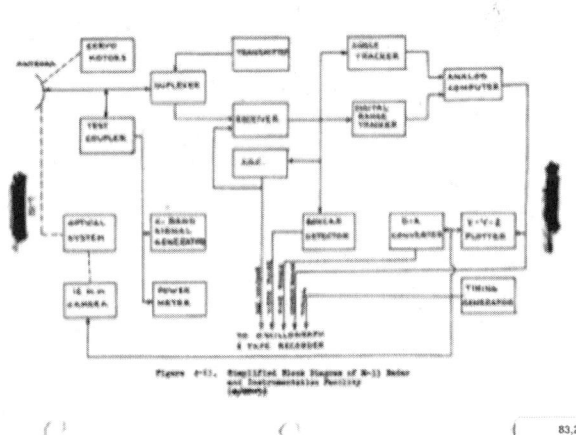

Radar cross section evaluation of the MiG-21

Through the CIA EG&G Special Projects past 18 months of RCS evaluations of the CIA A-12 stealth plane, the exploitation team realized the target signature information being fundamental in any assessment of the vulnerability of a weapons system to detection and tracking. Consequently, the Air Force element of the exploitation team sought both static and dynamic measurements of the MiG-21 using the Groom Lake ground based VHF band, S-band, and C-band radar systems with the Nike X-band radar providing the primary tracking role.

During the dynamic RCS evaluation, the MiG-21 flew a flight pattern at an altitude of 30,000 feet and at a speed of Mach .86. The aircraft maintained a wings-level attitude during straight portion of the flight. Lacking telemetry, the Special Projects exploitation team could not determine the aircraft attitude during turns. The Special Projects measurement facility made measurements at 170, 2,900, and 5,050 MHs using six basic radar subsystems with their antennas controlled by a specially designed Nike bullgear servo assembly and monitored from the control room of the EG&G Special Projects. Four of the radar systems participated in the measurement program. The VHF, S, and C-Band equipment gathered the reflectivity data while the X-Band system generated rands information and target bearings. The other radars slaved to the reference servo network controlled by the target bearing data from the Nike X-Band radar. The Special Projects exploitation team digitized and recorded on tape the range-gated video data received by each radar system for off-line computer processing. They did not use the VHF telemetry system that normally provided target pitch, heading, and roll data.

RCS climbing after take-off

MiG-21 Fishbed on the Groom Lake Pylon

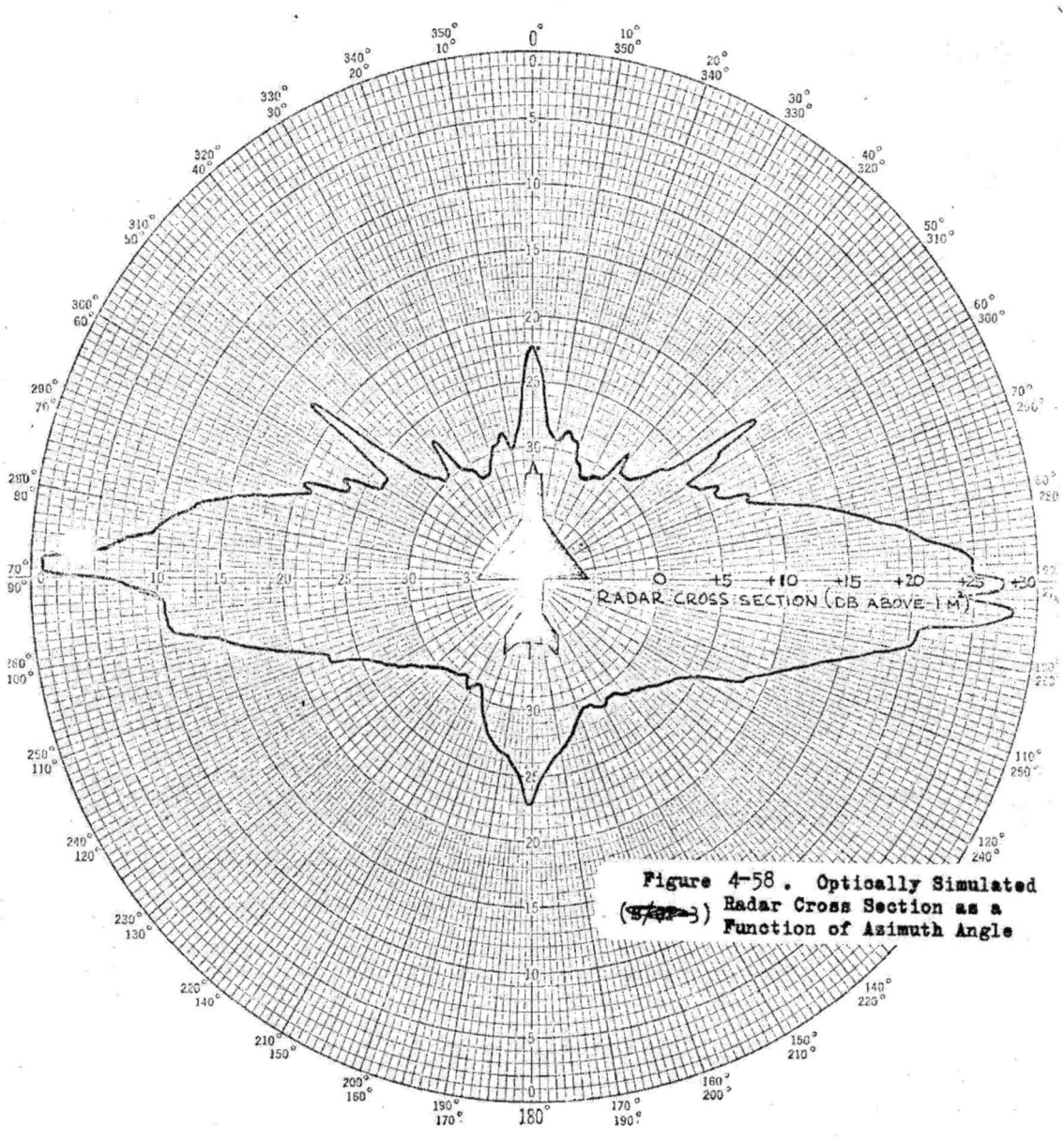

Typical RCS scan of a plane on the pylon at Groom lake by the Special Projects team.

High_Yo-Yo
1
2
3
1
2
3
2
1
Defender
3
2
3
1
Attacker

Maneuvers

High_G_Barrel_roll

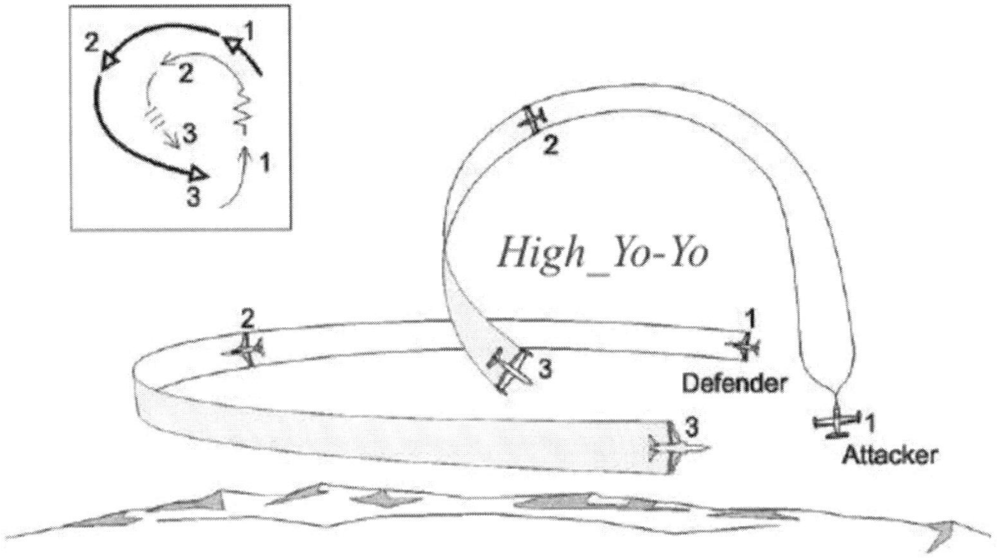

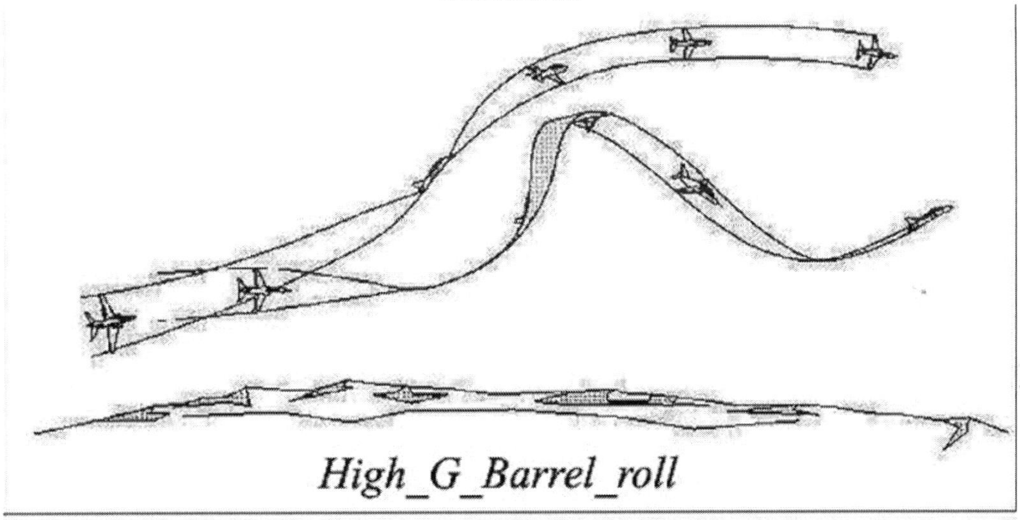

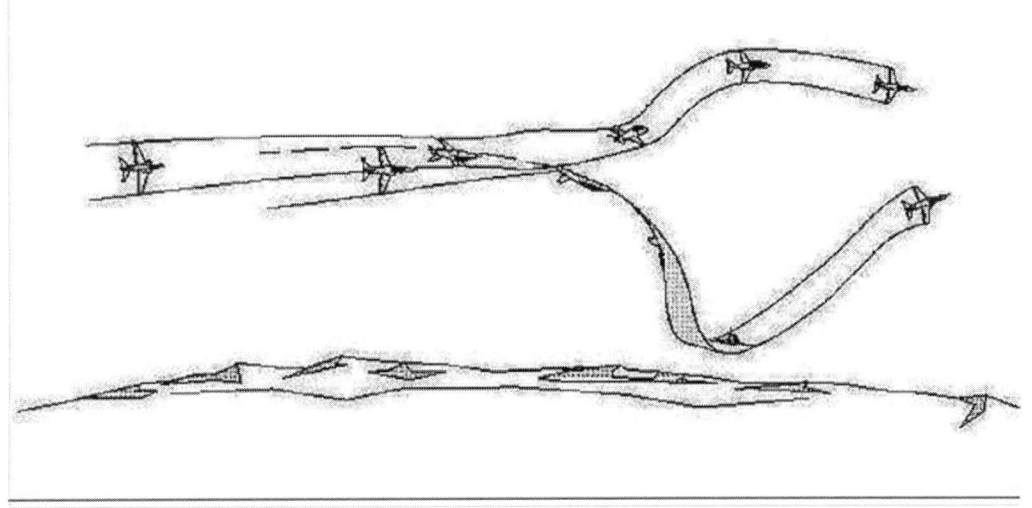

Ground visibility tests

CHAPTER 3 - HAVE DOUGHNUT Tactical Evaluations

The HAVE DOUGHNUT MiG-21F-13 delivered to the Israeli Defense Forces Air Force and loaned to the US for exploitation. Air Forces' Foreign Technology Division managed the exploitation in a joint Air Force/Navy technical and tactical evaluation at Area 51 with delivery on January 23, 1968.

The MiG-21 in Hangar Five at Groom Lake. Note the US Air Force decals. The plane was designated as a YF-110B.

The MiG-21 cockpit

Maj. Gerald D. Larson, TAC, observing Fred Cuthill preparing for a test flight, one of several in the MiG-21 (YF-110B) to evaluate aircraft performance, propulsion, subsystem, and design features, as well as measurements of infrared and radar signatures, engine modulation, and acoustic measurements.

HAVE DOUGHNUT exploitation team leaders

Major Fred J. Cuthill, chief of the Air Force Flight Center Branch, served as AFSC project officer to evaluate performance and handling qualities of the HAVE DOUGHNUT MiG-21.

Project officer Comdr. Thomas J. Cassidy, Jr., Naval Air Test and Evaluation Squadron 4 (VX-4) along with Lt Col Joe B. Jordan as TAC project officer for the Air Force TAC conducted tactical evaluation of the aircraft.

Later to be known as the Red Hats, the rest of the customer's exploitation team was Robert G. Ashcraft6–TAC, Gerald D. Larson–TAC (1137th SAS), William T. "Ted" Twinting–AFSC.

The pilot Familiarization Flight

On 8 February 1968, excitement ran high throughout the exploitation team at Groom Lake. Today, we, America, would fly the mystery plane, the crown jewel of USSR aviation. The flight-testing of the MiG-21[7] began with the Special Projects exploitation team performing

---

[6] In December 1977, the Air Force activated the 6513th Test Squadron Red Hats at Edwards AFB to support evaluation of foreign aircraft. Functional detail of the Red Hats remains classified

[7] The exploitation team flew the MiG-21 in natural finish with the number '90865' on the tail, and US

preflight on their radar systems and situating the Air Force and Navy personnel at the various control consoles while the Air Force and Navy participants towed the MiG-21 to its take off position at the end of the runway.[8] From their Special Projects position on the edge of the dry lakebed, the exploitation team participants first heard the start of the engine of the chase plane as it taxied to the runway, followed by the sounds of the participating plane starting up at the end of the runway. The screaming jet sound of the chase plane's engine and then the run up of the MiG-21 engine alerted the entire base that the flight of the Soviet MiG was about to begin. From that point, the small, secret base enjoyed viewing an aerial show as the exploitation team put the MiG-21 through a joint TAC and USN choreographed first flight of the MiG-21 Fishbed flying over Nevada to familiarize the pilot.

MiG-21, F-8E, F-4D

The February 8 flight lasted thirty minutes with the MiG-21 and F4D conducting two acceleration and two deceleration runs. Evaluation commenced on the ground evaluation with start, taxi, and takeoff where TAC found the MiG-21 engine response poor during taxiing and engine checks. Wheel brakes were fair and steering difficult with differential braking. Wheel brakes failed to hold during run-up at full power. Acceleration on takeoff and stabilator effectiveness on rotation was good, however, the landing gear failed to retract until the third recycle attempt with slight trim change during gear retraction, and the MiG-21 experienced a

---

national insignia on both sides of the nose, but not on the wings.

[8] To reduce the possibility of a ground mishap, the ground crew towed the MiG-21 to and from the active runway, also preventing unnecessary brake and tire wear associated with taxiing, a primary consideration since spares were not available.

slight sink during flap retraction. The pilot found the speed brakes poor and ineffective and aileron control very sensitive at low speed. Adverse yaw was very pronounced during low speed maneuvering. The afterburner refused to ignite until engine speed reached 100%, and the plane encountered airframe buffeting about 550 KIAS, which stopped with power reduction and deceleration. The Navy found the pilot strap-in cumbersome, required the assistance of a plane captain. Prestart cockpit checks were simple, but required concentration due to the cluttered switch panels and similarity of switches. The air operated canopy required approximately 10 seconds to close after actuation where it accomplished locking and seal pressurizing manually. The canopy required closing prior to moving the airplane. The Navy reported the pilot's visibility aft severely limited by the ejection seat headrest, windscreen, and armored glass combination.

With the Nike radar locked onto the MiG-21, the MiG-21, along with an Air Force a McDonnell Douglas F-4D Phantom II and Navy Vought-produced F-8E Crusader, climbed to 10,000 feet for an acceleration comparison, evaluation of afterburner and engine response, aircraft maneuvering qualities, slow speed handling characteristics, avionics and sight system analysis. Inside the Special Projects building, those monitoring the control consoles, radar, and data systems proceeded with a maneuvering flight of the MiG-21, watching as it entered a nose lightening and slight dig-in occurred at about 5.5 G. accompanied by high airspeed bleed off. The pilot reported it having medium to heavy stick forces as it approached stall where the plane experienced mild buffeting and wing rock. The pilot effected recovery at 140 KIAS as the left wing dropped. While accelerating from low speed, he reported the intake suck-in doors closing with a noticeable bang. Below 200 KIAS in the traffic pattern, the aircraft felt sensitive to controls.

From the TAC flight control console, a TAC officer directed the MiG-21 to ascend to 10,000 feet for an acceleration check with the F-4-D chase aircraft. During the run, the F-4 maintained a superior performance throughout (300 to 400 KCAS) in military power. Following that run, the planes conducted an afterburner acceleration check from 300 to 550 KCAS where the F-4E demonstrated that it could maintain a wing formation position with excess power available to separate from the MiG-21. At 550 KCAS, the MiG-21 terminated the acceleration because of severe buffeting.

A level flight deceleration with speed brakes followed that found the MiG-21 and F-4 equal in speed brake deceleration, however, at idle power, the F-4 decelerated more rapidly. Slow speed maneuvering required good piloting technique because of wing rolloff and adverse yaw characteristics of the MiG-21. The MiG pilot reported poor visibility through the forward windscreen with targets generally acquired at 3 to 5 NM range. The canopy also restricted rearward visibility so that the pilot saw only about one foot of each wing tip.

The MiG-21 pilot reported that selecting afterburner by moving the throttle into afterburner range from any position less than full military RPM delaying ignition until the engine accelerated to 100%. The F-4 chase pilot reported no apparent afterburner puff during AB operation at 10,000 feet and below and no apparent engine smoke at any time. Ground photos of the planes during maneuvers confirmed what the chase pilot reported.

The three planes grouped at 12,000 feet where the MiG-21 made a level CRT (Combat Rated Thrust) acceleration from 300 KIAS to 450 KIAS. The F-4D, flying along side, passed the MiG-21, however the F-8E, in trail, could not match the MiG-21's acceleration time of 50 seconds and dropped behind.

The planes made a level CRT acceleration at 12,000 feet from 300 KIAS to 550 KIAS with afterburners selected on signal from the Special Projects' control console manned by TAC and Navy mission controllers. The MiG-21 afterburner ignition delayed approximately 9 seconds with no afterburner puff observed. The F-4, flying along side passed the MiG-21, using half-modulated afterburner to maintain position. The F-8E MRT (military rated thrust) acceleration was also superior to the MiG-21. The MiG-21 deselected afterburner at 550 KIAS due to engine surging and/or airframe buffet with published Vmax (maximum speed) at this altitude as 595 KIAS.

The planes accomplished deceleration with speed brakes with similar F-4D and MiG-21 deceleration characteristics. The F-8E deceleration characteristics were superior to the MiG-21.

A 180-degree turn initiated at 550 KIAS and 5.5 g-The MiG-21 had adequate horizontal stabilizer available throughout the turn and experienced left wing dip and longitudinal control lightening at 5.2 g.

The Navy F-8E held a clear AIM-9D tone on the MiG-21 at 1.5 miles out to a 60 degree TC. The planes performed low speed turns to buffet at 250 KIAS (1.8.g left-2.0 g right) where the MiG-21 experienced light buffet at 2.2 g.

The planes performed straight and level, clean stalls at 78 percent RPM where the MiG-21 encountered light buffet at 185 KIAS, wing rock at 175 KIAS and left wing roll off at 150 KIAS. The MiG-21 lost approximately 500 to 800 feet in stall recovery due partially to slow engine acceleration. The MiG recovered in a slightly unbalanced flight and rolled off in the direction of adverse yaw at 175 KIAS. The pilot of the MiG-21 felt comfortable at 300 to 500 KIAS, but below 200 KIAS, he reported, "The airplane feels squirrely." During the later debrief, the pilot stated his feeling that only a well-experienced pilot should attempt scissors, reversals, etc., in the MiG-21 to maintain completely coordinated control in this area, and to avoid unstable flight and subsequent departure.

Return to base was uneventful with a low approach followed by a final landing. The pilot used his speed brakes to maintain high engine RPM during the approach, extending his landing gear and flaps to 250 KIAS. In the landing configuration, longitudinal control lightening occurred at 140 KIAS followed by wing roll-off, flying the base leg at 200 KIAS with 170 KIAS on final with normal touch down. The drag chute actuated after touch down at 140 KIAS and deployed at 110 KIAS. The pilot selected two-wheel braking prior to turning off the runway. Engine shutdown was normal. During debrief both the Air Force and the Navy found the MiG-21 extremely small and difficult to see. It left no apparent smoke trail at any power setting.

On 11 February, the exploitation teams conducted a second familiarization flight that lasted 35 minutes with the MiG-21 and F-4 conducting two acceleration runs while evaluating handling characteristics, avionics equipment, and climb comparison.

Tactical evaluation of the MiG-21

Moving into the tactical phase of exploitation, the National Air and Space Intelligence Center, along with the EG&G Special Project team, conducted choreographed USAF and USN combat flights from the Special Projects mission consoles. There, they evaluated the effectiveness of existing tactical maneuvers in an air-to-air environment, optimizing existing tactics, and developing new tactics while evaluating the design, performance, and operational characteristics of the MiG-21

For offensive and defensive evaluation, TAC evaluation aircraft flew 35 of the 102 total sorties on the MiG-21 with the F-4C/D/E, F-105D/F, F-111A, F-100D, F-104D, F-5A. The RF-

101, RF-4C, and B-66 flew defensive evaluations only. The Special Projects personnel supported the exploitation team in acquiring maximum documentation during the data acquisition phase. In addition, the safety chase aircraft, participating aircraft, and ground monitor utilized gun camera and external pod-mounted 16 mm canon Scopic motion picture cameras when possible on selected flights. To record cockpit conditions during each flight cockpit, the exploitation team installed over-the-shoulder cameras in the MiG-21. The MiG-21 used voice tape recorders, safety chase, and participating aircrews to document each tactical air situation and development. Ground monitors recorded the UHF communication during comparative flight evaluations. The exploitation team recorded and summarized briefings and debriefings daily. All participating aircrews completed flight data cards during or immediately after each flight with significant events and pilot qualitative comments noted.

MiG-21 FISHBED exploitation maneuvering at Area 51

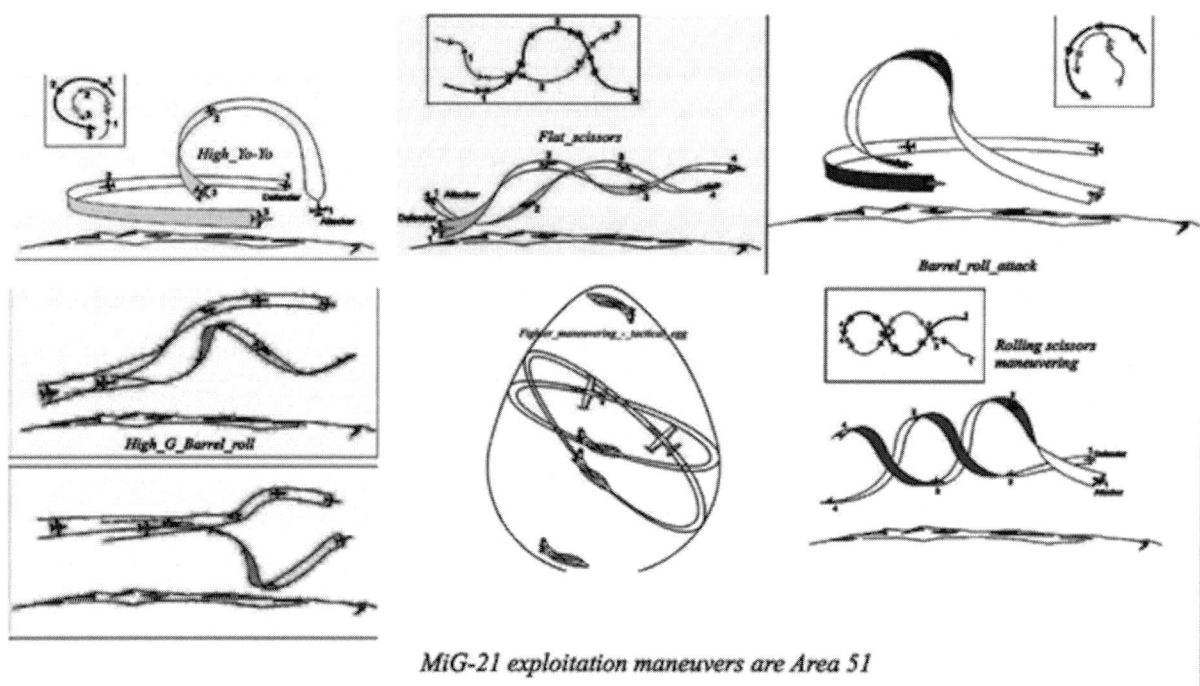

MiG-21 exploitation maneuvers are Area 51

US planes evaluated against MiGs at Groom Lake

Navy F-4 Phantom II

Navy F-4J EI

Navy A-4F Skyhawk

F-5N

F-101

B-66

F-100D

F-104

F-205D

F-111A

Navy F-4 Phantom II

Navy F-4J EI

Navy A-4F Skyhawk

Navy F-8E Crusader

Air Defense Command F-106

Strategic Air Command B-52

Navy A-7a Corsair II

Joint Air Force and Navy technical evaluation flights on the MiG-21

During evaluation of the MiG-21 infrared radiation measurements, the joint Air Force-Navy exploitation team used an APGC ATR-1/T-39 airborne radiometer system provided by the China Lake NWC with the MiG in cruise, military, and afterburner modes. The T-39 measuring aircraft furnished by Wright Patterson AFB and a chase plane equipped with a standard IFF transponder took formation in a racetrack patter about 50 nautical miles long. For acquisition, the MiG-21 positioned about one mile in front of and slightly above the radiant-intensity measurement aircraft with the chase aircraft on the wing of the MiG. The measuring aircraft with the Navy's infrared radiometric equipment and IF missile seekers established infrared tracking of the MiG with the ATR-1 radiometer at the same time that ranging equipment locked on to the IFF. The pilot of the MiG provided pertinent performance data by UHF communication on the MiG's pertinent performance data through various power settings and maneuvers. Among the many radiation characteristics of the FISHBED E aircraft, the more interesting data revealed no significant increase in values during altitude increase or at extended slant ranges.

The exploitation team successfully conducted a ground-based IR measurement program against the MiF-21 FISHBED E both on the ground and in the air. They did so using an infrared television, two infrared radiometers, A CVR (circular variable filter) spectrometer, and a movie camera collected data during engine run-ups with the aircraft sitting at the end of the runway before takeoffs, during takeoffs and land, and during flyovers. They recovered useful data for infrared search, warning receivers, and similar equipment. The IRTV sensor successfully detected the FISHBED in tail aspect and without afterburner at 40 KM. They found the exhaust plume of the FISHBED E significantly smaller than the exhaust plumes of F-4 and F-8 aircraft in the same wavelength region.

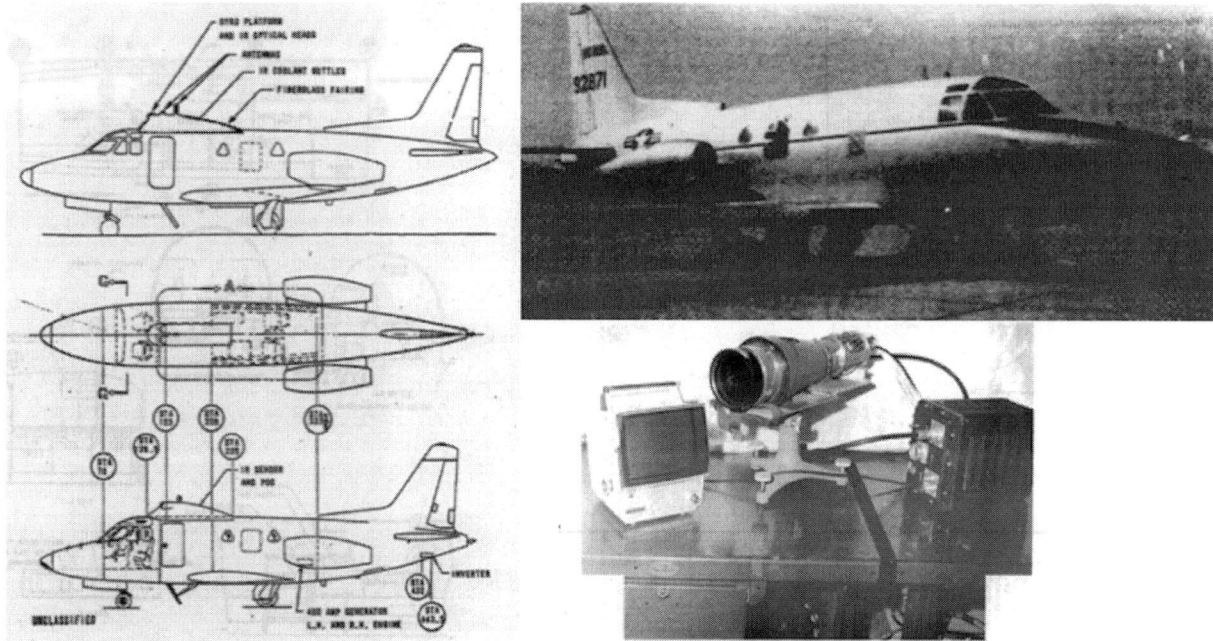

APGC ATR-1/T-39 airborne radiometer system

The Wright Patterson Air Force Base T-39 twinjet aircraft serving as an airborne platform for the infrared color-wheel radiometers and for the missile-tracking display instrument provided by NWC proved to be a very reliable platform able to stay aloft for three hours and fly to an altitude of 45,000 feet at a maximum speed of Mach .8. To prepare for a test flight, the exploitation team removed the T-39 starboard passenger seats and replaced the original starboard escape hatch by either of two modified by NWC, one to accommodate a turret designed for use with the radiometers and another to accommodate a fiberglass turret used with the missile-tracking-display units. They fitted each turret with infrared-transmitting windows mounted on gimbals to permit wide-angle manual tracking. Snap-pins insured quick installation and proper alignment of the instruments in the turret. The exploitation team placed the radiometer power supply and a signal-monitoring oscilloscope on the deck and secured it to the bulkhead by a metal snap-band. A nearby panel provided 400-cycle aircraft power, and transferred warm air to the turret windows from aircraft heating ducts via a flexible rubber tube to prevent the formation of frost.

The exploitation of the IR detection of the FISHBED E found it more susceptible than either the P-4 or P-8 aircraft to a tail attack by IR air-to-air or ground-to-air missiles operating in the 2 to 4 micron region. The smaller intense plume of the FISHBED E provided better guidance characteristics to the missile since it more closely resembled a point source. A missile would guide better on the smaller FISHBED E plum, making it much more susceptible to contact or influence fused TR missiles than the P-4 and P-8 aircraft. A smaller miss distance to the missile would better insure the MiG-21 being within lethal range when the fuse activation/warhead detonation occurred. Consequently, the FISHBED E required a more intense flare than either a P-4 of P-8 aircraft to successfully flare decoy an IE missile where the spatial filtering provided by the reticle in an IR missile guidance head specifically rejected extended area (i.e., clouds, etc.) targets. A missile launched against a FISHBED B would see a small area plume and would be less apt to spatially reject it in favor of a point source flare.

AIM guided missiles

The AIM-7 missile was a supersonic air-to-air homing missile substituted for the ATOLL missile during HAVE DOUGHNUT. The launching airplane supplied guidance by the transmitted CW signal. To lock-on and track the target by means of proportional navigation, the missile compared the CW signal received from the launching airplane with CW signal reflected from the target.

The A1M-9B/D missiles were supersonic air-to-air homing missiles employing passive infrared target radiation for guidance also substituted for the ATOLL missile. The missiles use the infrared energy emitted by the target to lock-on and track the target. Target detection provided an audible tone to the aircrew when achieving target detection.

Radar homing and warning set AN/APR-25

The AN/APR-25 is a passive ECM (Electronic Countermeasures) set and consists of warning receiver, threat indicators, and an audio system. The radar homing and warning set indicated threat bearing relative to airplane heading, relative signal strength, and the type of emitting activity. Threat signals were discriminated from non-threat signals by radio frequency signal strength, pulse repetition frequency, and antenna scan rate. Threats displayed on the strobe display scopes and on threat indicators in each cockpit.

The MiG-21 size and shape looked very similar to an A-4. The A-4, A-6, A-7 possessed sufficient maneuverability in an initial break turn to thwart a MiG-21 attack. However, resulting energy loss allowed MiG-21 to reattack with comparative ease if the A-4/6/7 stayed in the fight at slow speed The MiG-21 pilot could choose to engage or disengage the A-4, A-6, A-7 aircraft at will. The A-4 and A-6 could reverse and obtain a quick snapshot if the MiG-21 overshot close in. The A-7A did not appear to have the necessary thrust to complete this nose high maneuver.

The MiG-21 30 mm cannon

The MiG-21 30 mm cannon were 100 percent reliable. The gun fired on five different sorties. On each sortie, ten HEI (high explosive incendiary) rounds were loaded and fired out in a single burst in near one g flight at a ground target with no jams or failures to fire encountered. The muzzle flash was readily visible at one mile in daylight.

The exploitation team demonstrated the lethality of the 30 mm cannon in a ground attack flight conducted against a standard.

The exploitation team demonstrated the lethality of the 30 mm cannon during a simulated ground attack flight. The target for the strafing attack was a standard, US manufactured bulldozer. The weapon rendered the bulldozer inoperative and irreparable after hitting it with two1 round of 30 mm HEI ammunition. The 20 mm cannon used by US tactical fighters failed to have caused a comparable degree of damage.

Weapon targets for MiGs at Area 51

The MiG-21 Armament and Fire Control System

The exploitation of the FISHBED E Armament and Fire Control System found that it suffered from its forward visibility degrading the overall effectiveness of the FISHBED E weapons system, requiring the pilot to retain visual contact of the target to effectively employ the tracking index, fire control systems, and armament. The sight combining glass, the bulletproof glass slab, and the forward windscreen caused targets to be frequently unobserved, or not visually spotted.

While evaluating switchology, during air-to-ground attack flight, the pilot generally had enough time to position switches as required, but found the air-to-air switchology requirements initially to set up the systems more difficult and requiring excessive pilot actions. It was possible, however, to convert from a missile attack to a gun attack with one switching movement.

The armament of the MiG was surprising. The HR-30, 30 mm cannon was limited to a capacity of 60 rounds with an estimated gun rate of 850 rounds/minute, which provided for a total firing of 4.2 seconds. Estimated muzzle velocity was about 2,560 feet/second. During gunfire, pipper jitter was excessive, about 20 mils, and tracking correction during gunfire was not possible. The participants saw the muzzle flash during daytime conditions from a range of 3 miles.

The exploitation team used a bulldozer as a target during the tactical phase of the exploitation. The bulldozer was in operational condition before this attack, although the exploitation team removed the blade. An estimated two rounds of HEI impacted the vehicle and rendered it irreparable. The evaluations encountered no cannon malfunctions during the cannon firing missions.

Manual ranging of the gunsight could not perform smoothly and precisely. System hysteresis-and friction made it virtually impossible to prevent over control of the sight reticle diameter size with the throttle twist grip, pipper jitter during cannon firing was in excess of 20 mils.

Gyro drift when tracking air targets was excessive. G loads greater than 2.5 caused the sight reticle to drift to a point near the bottom of the sight combining glass. At very high g loads, the sight reticle disappeared entirely.

Sight electrical paged function was sluggish and slow to respond. During air-to-air tracking, it was necessary to hold the electrical cage button (on the stick grip) until radar lock-on occurred. The electrical page button was poorly positioned and difficult to actuate when preparing to fire the gun.

FISHBED over-the-nose visibility restrictions limited the useful mil depression to 94 mils. Large lead angles during air-to-ground attacks with bombs, gun, or rockets were not available. It was not possible to depress the gunsight in the gun mode of operation as may be required for ground attack at long slant ranges.

The exploitation did find some desirable characteristics incorporated in the MiG-21F-13 FISHBED C/E Weapon System that included ease of system operation, simple cockpit procedures, and minimum system monitoring enhance pilot performance during a tactical engagement.

The small frontal area provided a low probability of visual or radar detection of the MiG-21 in a head/tail-on aspect. The MiG-21 pilot could use the quality to his advantage for reduced detection during patrol or attack. After initial visual detection of the MiG-21, it was necessary to "padlock" or remain visually fixed on the aircraft to prevent losing contact. US tactical aircraft of comparable size were the F-104 and F-5.

Operational weight of the FISHBED, configured with 60 rounds of 30 mm ammunition and 2 ATOLL missiles is 16,250 pounds. Although not demonstrated, it was possible to operate the MiG-21 weapon system from soft runways, i.e., snow, dirt, sod, etc. Tire consumption rate was comparatively low and 50 landings were normally available from main gear tires. During this evaluation, one set of tires accumulated 53 landings.

The MiG-21's head up display provided the pilot with target radar lock-on, target in range, and over-g condition for missile launch, and target breakaway (minimum range). Although lacking in sophistication, the presentation provided the MiG-21 pilot with required information in a manner, which was simple and effective.

TAC inventory vs. MiG-21F FISHBED flight test overview

The joint TAC, USN, CIA EG&G Special Projects, and other government agencies' tactical exploitation phase flew 40 of the available 52 days with eight days cancelled due to weather and four days due to maintenance. Of the 134 sorties scheduled, the tactical phase flew 102 sorties, canceling 21 sorties due to weather and 11 due to maintenance. The flights consisted of one acceptance flight, 1 photo, 1 RCS, 9 IR, 2 SAC, 4 ADC, 26 performance, stability, & control, 33 TAC tactical and 25 USN tactical. Some were solely Air Force or Navy and some were joint exploitation and evaluation flights.

Various customers in the exploitation desired and conducted their own tests using their own airplanes and seeking solutions that differed from the other participants. Therefore, where there might be duplication, it is because this book covers each participant's interest, tactical evaluation, and conclusions.

During the exploitation of the MiG-21, the USAF and USN evaluated the aircraft performance, aircraft stability and control, armament, and cockpit environment again the F-4C/D/E, F-105D/F, F-111A, F-100D, F-104D, F-5N, RF-101 (defensive only), RF-4C, and B-66 (defensive only).

Summary of Sorties			Maintenance	
USN Tactical	25		MiG-21 Aircraft at Test Site	75 days
USAF Tactical	33			
USAF Performance, Stability & Control	26		Reassembly	15 days
Air Defense Command	4			
Strategic Air Command	2		Bell Mouth Ground Run	3 days
Infra Red	9		Disassembly	5 days
Radar Cross Section	1			
Photo	1			

Emphasis on what was learned from matching the F-4 against the MiG-21

From a combat spread of 1 mile, the F-4 section meeting the MiG-21 head-on immediately selected afterburner, accelerated, and separated to effect a VID maneuver or commence loose deuce maneuvering. The maneuver forced the MiG-21 to pick one F-4. At that point, the engaged F-4 completed a head-on pass followed by an oblique loop as described in the one-on-one tactics. This freed the second F-4 to maneuver and immediately press for the offensive. The F-4 making the head-on passes kept the MiG-21 engaged while the free F-4 maneuvered into a missile launch position. It was essential that the "free" F-4 maneuver rapidly in a different plane to strive for a rear quarter attack. If the MiG-21 switched to the free F-4 during the engagement, the F-4s also switched positions making the previously engaged F-4 the free F-4.

When both F-4s reached an astern position on the MiG-21, the basic tactics previously described applied to the attacking F-4. While one F-4 committed to an attack the other positioned himself out of the plane of the maneuver, preferably in a high cover position, and be ready to conduct a slashing attack. Since the MiG-21 had a high rate of turn and small turn radius, an F-4 high Yo-Yo easily resulted in a head-on pass coming down from the apex of the Yo-Yo.

The exploitation team had to be proficient in ACM and familiar with the maneuvers described in the F-4 Tactical Manual that described defensive maneuvering by the attacked F-4 in one-on-one tactics. Early separation in the vertical by the free F-4 provided mutual support was necessary to gain a missile launch position and sandwich the attacker. If the MiG-21 switched and attacked the higher F-4, the high F-4 had to break down into the attack, inform his exploitation teammate of the switch, direct the exploitation teammate to ease turn, and execute on oblique loop. Passing through the vertical, the pilot sighted his exploitation teammate passing underneath on a near reciprocal heading, with the MiG-21 pursuing or remaining high and switching his attack. If the MiG-21 continued pursuing the low F-4, the low F-4 continued separating while the high F-4 completed the loop behind the MiG-21. If the MiG-21 remained high, the high F-4 called the switch and directed the free F-4 to execute an oblique loop. The engaged F-4 generated a large overshoot then dove for sufficient separation to allow a reversal for a head-on pass, while the free F-4 maneuvered for the kill.

The same basic tactics applied if encountering multiple bogies. Section integrity, lookout doctrine, and mutual support were mandatory. The more complicated the tactical situation, the tighter the section maneuvering became. Split plane, vertical maneuvering had to continue once the attacked bogey aggressively maneuvered defensively, or attacks the section.

To eliminate tracking problems through MBC after initial detection in pulse Doppler, F-4J aircrews initiated rapid relock to the pulse mode, initiating rapid relock as early as possible, contingent on the following factors:

Delay rapid relock on a suspected MiG-21 target until range well inside the expected detection range of the MiG-21 by the pulse radar system.

Differential altitude. Target look down angle eliminated prior to initiating rapid relock.

Intercept geometry. Stopping target drift prior to a rapid relock attempt, unless a rapidly changing Vc indicated that target contact was lost in main beam clutter. This case made mandatory an immediate attempt to relock in pulse mode.

During evaluating target maneuverability, pulse relock eliminated the possible loss of radar illumination as a maneuvering target enters the main beam clutter notch. Ideally, the flight accomplished pulse acquisition and acknowledgement prior to 20 nautical miles during a VID maneuver.

Continual center of the split elevation strobe and expanded velocity display was mandatory prior to rapid relock to Pulse. Periodic use of the pulse mode made during BARCAP/TAR-CAP operations to offset the difficulty in detecting low Vc targets, i.e., Vc lower than F-4J TAS.

The extensive turbine modulation of the MiG-21 was a possible aid to pulse Doppler detection of low Vc MiG-21 targets. Aircrews had to be aware of the erroneous Vc information presented, but use it to assist in rapid relock.

Whenever possible, avoid assignment of co-channel and adjacent channel APG-59 radars to the same section or CAP station. When unavoidable, exclusive use of pulse Doppler mode was not reducing the effect of mutual interference.

F-4s equipped with pulse-only radars were aware of the short-range contacts probable during overland operations at current BARCAP/TARCAP altitudes. In an area of known MiG activity, visual search, and random heading changes aimed at thwarting a GCI-controlled MiG-21 intercept took precedence over radar search.

Due to pilot field of view restrictions in the F-4 and the small size of the MiG-21, it was essential that RIO's utilize "padlock" lookout technique while engaged in ACM. The pilot made no attempt at radar acquisition until the MiG-21 was within approximately 45 degrees of the nose and confirmed visual contact to eliminate unnecessary position calls that might block out UHF transmissions from the accompanying F-4.

RIOs were thoroughly familiar with all phases of ACM to assess an enemy maneuver and provide proper directive commentary if the pilot lost visual contact for a protracted period. During actuation of PLM, the RIO switched his attention to the radarscope and assisted the pilot in determining the proper direction of range gate sweep, based on the relative positions of the target and altitude line. After lock-on, the RIO verified valid target track by noting correct Vc, illumination of range track light and AIM dot, and elevation strobe deflection. He notified the pilot and called the range to the target. On obtaining a false Lock-on, he broke lock and told the pilot to re-acquire. Timely and accurate information from the RIO reduced the possibility of launching an AIM-7 out of envelope.

Due to rapidly changing target aspects, the pilots used the pulse Doppler mode only as a secondary mode of operation during ACM. If conditions warrant, they employed automatic acquisition. However, they anticipated spurious automatic lock-on on other airplanes in the area or beyond visual range.

In an area where a number of friendly I-band emitters were present, visual and radar search took precedence over APR-25 indications for evidence of.MiG-21 activity. Under no circumstances did they ignore an APR-25 I-band track indication of an unknown source.

Conversely, attention to APR-25 search did not detract from other means of detecting the presence of MiG-21, i.e., visual and radar search.

A MiG-21 was most difficult to acquire visually in excess of two nautical miles. For this reason, a rigid section lookout doctrine was mandatory if the section expected to operate successfully in a hostile area. Section loose deuce maneuvering, separated in the vertical plane, was most effective formation for converting a defensive situation to the offensive.

Attack airplane employing Tactical Manual defensive maneuvers against the MiG-21 proved effective. The exploitation team validated all maneuvers in one-on-one engagements and stressed lookout doctrine. Maintaining a tactical section provided mutual support and to enhance lookout doctrine.

USN MiG-21 findings

The Navy concluded that the MiG-21 was extremely difficult to visually detect and keep track of in the ACM environment. The MiG-21 had a definite tactical advantage due to its small size, and was a highly maneuverable airplane capable of high g, low speed flight with a Mach 2 capability at high altitude.

- A clean MiG-21 encountered heavy airframe buffet at .96 IMN below 16,000 feet, however, its maneuvering flight characteristics were limited by high stick forces at speeds above 510 KIAS below 16,000 feet.

- The MiG-21F's turning ability was impressive due to low wing loading and high thrust to weight ratio. Flown to maximum performance, it out turned an F-4 and F-8 series airplane in a close-in turning engagement.
- The MiG-21 zoom performance up to 25,000 feet was inferior to the F-4 fighter configured without a centerline tank.
- The MiG-21's zoom performance was comparable to an F-4 fighter configured with a centerline tank.
- The MiG-21 zoom performance was comparable to the F-8 below 25,000 feet.
- The MiG-21 on station ACM time was comparable to the F-4/F-8 series airplanes with a similar percentage of total fuel on board.
- The MiG-21 gun sight system tested had limitations in the ACM environment.
- The MiG-21 airspeed bleed-off in a high q turn was rapid below 400 KIAS.
- The MiG-21's total performance degraded only slightly with the installation of Sidewinder type missiles.
- The MiG-21's engine acceleration was slow from 85 percent to 100 percent.
- The MiG-21 leaves little or no smoke trail at military or afterburner power settings.
- The MiG-21 cockpit visibility suffered serious degrading through the forward windscreen, below the canopy rails and in a 50-degree cone aft.
- The MiG-21 30 MM cannon were effective and reliable.
- The MiG-21 was an extremely vulnerable airplane.
- The MiG-21 was easy to maintain.
- The MiG-21 sortie rate was high.
- The MiG-21 was capable of being recycled in 30 minutes.
- The F-4/F-8 series airplanes have a tactical disadvantage in the ACM environment because of their large size and prominent smoke trails.
- F-4/F-8 series airplanes were capable of exceeding the MIT-21's q limit at low altitude.
- F-4/F-8 series airplanes have better CRT acceleration performance than the MiG-21 below 1.2 IMN at low and medium altitudes.

RECOMMENDATIONS

The Commander Operational Test and Evaluation Force recommended that:
- Navy fighter airplanes engage a MiG 21 in section.
- Maintain section integrity throughout the engagement for mutual support.
- Strict lookout be maintained at all times and the engaged airplane utilize "padlock" lookout technique.
- In a threat area, weave and vary headings along the base course.
- Force all engagements to low altitudes at high speeds.
- Determine the MiG-21 pilot's ability early in the engagement.
- A close-in, slow speed engagement be avoided if the MiG-21 was flown at or near maximum performance.
- Orient offensive maneuvering toward exploiting MiG-21 weaknesses rather than rushing for a quick kill.
- Keep an attacking MiG-21 at high TC.
- Aircrews be aware of situations or conditions that warranted disengaging from a MiG-21 encounter.
- Employ only sound, proven tactics.
- Practice ACM under controlled conditions against small airplanes with low wing

loading, e.g., A-4F, E-5.
- Aircrews be thoroughly familiar with and aware of their weapons system limitations.
- Exploit the limitations of the enemy.
- Whenever possible, radar Intercept and air combat conduct training over land.
- Make AIM-7E attacks on a maneuvering MiG-21 in the pulse mode of the APG-59 radar.

The USN concluded that the aircraft was easy to fly with no dangerous characteristics. Design of the vehicle avoided complexity occurred whenever possible. Particularly noteworthy was the fact that the plane employed no stability augmentation.

The acceleration and thrust-limited turning performance of the airplane, although less than predicted, was good throughout the flight envelope.

Basic airplane stability, except lateral-directional damping, was good. The airplane exhibited excellent lift-limited maneuvering characteristics in terms of both the available load factor and handling qualities near the stall. Roll rates and roll response were good through the flight envelope.

In turbulent conditions, the aircraft was not an acceptable platform for weapons delivery or instrument flying because of weak lateral-directional damping combined with slow engine response.

At low altitude in the transonic region (0.96 to 1.15 IMN), the airplane vibrated to such an extent as to preclude its use as a weapons delivery platform. The intensity of the vibration at a given Mach number increased with decreasing altitude; below 15,000 feet, the cockpit instruments vibrated to the point where they were almost completely blurred.

Engine response was poor; the engine accelerated slowly even at high power settings. The poor engine response precluded precise formation flying.

The cockpit design was antiquated. It was not possible to enter the cockpit with any degree of urgency because of the time-consuming tasks associated with donning the parachute harness and hooking up the necessary personnel leads. Forward visibility was poor. Labeling of the switches in the cockpit was inconsistent; the labels on the right side were above the switches, whereas the labels on the left side were below.

The MiG-21 looked very similar to an A-4 airplane in size and shape. When both airplanes were in the same area, all pilots commented on the difficulty in determining one from the other.

The A-4F, A-6A, A-7A airplanes possessed sufficient maneuverability in an initial break turn to thwart a MiG-21 attack. During every engagement where the MiG-21 attacked, fixed wing, heavier than air attack planes (VA) (attack) forced the MiG-21 into a high Yo-Yo maneuver or an overshoot. VA airplanes, while executing the break turn, lost considerable energy and g available, allowing the MiG-21 to reattack with comparative ease if the VA airplane elected to remain in the fight at slower speeds than the MiG-21.

VA airplanes had no control over the MiG-21's ability to disengage at any time throughout an engagement. The MiG-21's higher thrust to weight ratio allowed the MiG-21 pilot the option of continuing the engagement or disengaging at his discretion. The MiG pilots forced the VA airplanes to counter his offensive maneuvers as long as the planes remained engaged.

A-4F and A-6A airplanes demonstrated an ability to reverse and obtain a quick snapshot, if the MiG-21 overshot close in at a high TCA (50° or more). When the MiG-21 overshot close in, the A-4 and A-6, using a maximum performance rudder reversal, were able to attain a position behind the MiG-21 within the AIM-9D launch envelope. Speeds for the A-4 and A-6 after completing this maneuver were very low (about 130 KIAS). The A-7A did not appear to have the thrust available necessary to complete this nose high maneuver.

Results and Exploitation team Discussion MiG-21

The Air Force Flight Test Center completed a ground test, performance, and stability evaluation of the MiG-21 F-13 FISHBED E, finding it a simple and highly reliable Mach 2 airplane.

In general, during the technical exploitation the exploitation team members found the MiG-21F-13 cockpit reflected the Soviet philosophy of engineering simplicity. Poor functional grouping of switches, controls, instruments, and warning lights gave the cockpit a cluttered appearance. The poor grouping of switches and controls required close pilot attention when requiring some cockpit action. However, overall design simplicity of aircraft systems generally required little pilot monitoring or control.

No major maintenance malfunctions occurred during the 102 MiG-21 flights. The team changed the tires, wheel brakes, and engine oil filter after approximately 50 flights. Three minor engine EGT (Exhaust Gas Temperature) system problems occurred. The team noted one small hydraulic leak. Some difficulty was experienced during the initial flights with landing gear retraction, but did not affect airplane availability. The canopy operating system was potentially weak, but careful actuation of the canopy controls minimized failure.

The MiG-21 required minimal ground-handling equipment for over-the-wing refueling and re-servicing of gasoline, lubricants, gaseous oxygen, and high-pressure air. Filler ports and access panels were readily accessible and aircraft turn-round time frequently was 30 minutes. Battery starts were possible; however, the exploitation team normally used an external power source during this evaluation.

MiG-21 servicing requirements were minimal. A crew of six men serviced and maintained the MiG-21, often completing servicing between flights in 30 minutes without difficulty. All servicing receptacles were readily accessible through individual access panels.

The MiG-21 systems were unsophisticated and designed for high reliability. No complicated servicing equipment was required. The MiG-21 had a self-contained electrical starting unit. The main and booster hydraulic systems pressurized to a maximum working pressure of 3100 psi, but normally operated at 1200 to 1400 psi as compared to constant high pressures in US airplanes. The pneumatic system was a ground charged, highly reliable system rated at 1800 psi. The fuel system was gravity filled and pressurized by sixth stage engine compressor air.

One boost pump supplied fuel directly to the fuel control. The 28 V electrical system contained a battery, starter-generator, and an inverter. Emergency electrical power was available from two batteries.

The MiG-21 was corrosion free. The skin's clear lacquer coating of the airplane did not crack, peel, or deteriorate. The FTD, US AFSC, placed this substance under analysis.

The MiG-21 cockpit noise level was low. Compressor air and heat exchanging provided cockpit pressurization and air conditioning. The automatic cockpit temperature control preset prior to take off could not reset in flight, however, manual air temperature control was available.

No fog or snow blew into the cockpit through the system on any flight. The cockpit noise level was much lower than the F-4 or F-8 airplanes.

The MiG-21 pilot seat positioning appeared to enhance pilot g tolerance. In the cockpit, the pilot sat with knees raised and his legs pointed more forward than down. As a result, the pilot's g tolerance appeared to be raised approximately one g.

The MiG-21 had poor cockpit visibility. The combination of a bulletproof glass plate, the gunsight combining glass, and the canopy restricted visibility through the forward windscreen to 3 to 5 nautical miles against F-4/F-8 sized targets. Visibility through the forward side panels and the remainder of the forward hinged, clamshell canopy restricted the pilot's head movements, resulting in a 50-degree blind cone to the rear. The canopy rails were much higher than in US airplanes and limited look down at the three and 9 o'clock positions to 20 degrees. "S" turning and looking through the forward side panels obtained excellent forward visibility in the MiG-21.

The project participants classed the MiG-21 cockpit layout and seat mechanization as generally poor, requiring a ladder to gain access to the cockpit.

Prior to pilot entry, maintenance personnel manually adjusted rudder bars. The pilot stepped on the seat, which contained the parachute; supported himself on the canopy rails; and carefully positioned his feet on the rudder bars. He then lowered himself into the seat. The pilot took great care to position his feet on the rudder bars because of the limited space between the leg restraint mechanism, center pedestal, and the lower instrument panel.

Seat comfort was marginal because of the parachute harness back strap arrangement, which the exploitation team alleviated somewhat on some flights by putting a foam rubber cushion between the harness and the pilot's back. The legs and buttocks positioned on the same level, which reduced the tendency for blood to pool in the lower body areas as g forces were applied. The pilot accomplished up and down seat adjustment by an electric actuator. Canopy/head clearance restricted head movement. Seat adjustment accomplished by an electrical actuator, which moved the seat up and down. Seat positioning optimized pilot body posture so that the pilot more easily tolerated high g loads.

The canopy was pneumatically operated by controls within the cockpit and externally accessible from the left forward nose section. The pilot positioned two levers in the cockpit to close and lock the canopy. There was no warning light to indicate a canopy-unlocked condition. This required care when opening the canopy so as not to apply pneumatic pressure to the actuator before the locking mechanism had fully released. On one occasion, improper opening technique caused the canopy to snap open forcefully and disengage at the forward hinge point.

The MiG-21 required closing the canopy when taxiing. Limited over-the-nose vision and reduced acuity through the forward windscreen resulted in poor visibility when tracking. The narrow canopy restricted head movement. Ejection triggered on each armrest appeared easy to operate and were readily accessible.

Donning the parachute and integral seat restraint harness required one to two minutes. Each leg strap on the seat-type parachute positioned over the leg and threaded through a harness loop and seat pan slot at the rear of the seat, then into a central harness connector. Finally, the pilot snapped the right shoulder strap into the connector and attached the oxygen, anti-g suit, and communications leads. The personnel lead group, although bulky, did not restrict pilot movement or cause discomfort. A ratchet handle located on the right side of the seat allowed the pilot to tighten the harness and restraint mechanism to a high tension. Shoulder harness slack was adjusted by a release/locking lever located on the left side of the pilot seat.

The MiG-21 cockpit switches were considered poorly located. With slight slack in the shoulder harness, the pilot could actuate all switches and controls. With the shoulder harness locked in the fully retracted position, the pilot had some difficulty reaching the forward left and right extremities; i.e., landing gear panel and indicator light dimmer control.

Identifying placards positioned above each switch located the switches on the right vertical console. This inconsistency was confusing to an inexperienced MiG-21 pilot and caused identification difficulty. Guards and cover's for switches and buttons were good. Armament switches, controls, and monitoring lights for bombs, rockets, cannon, and missiles were located at random throughout the cockpit. Despite this scattered switch location, very little pilot action was required to set up the desired armament. When converting from a missile to cannon attack, the pilot had to reposition the Missile—Cannon switch to "cannon," and uncage the sight cage lever, the latter accomplished by alternate use of the electrical cage function.

The MiG-21's instruments were poorly grouped and located. Pilot crosscheck required total panel scan instead of localized scanning. The instrument panel contained Mach meter, vertical speed, and turn indicators on the right half of the instrument panel; and the attitude indicator, airspeed, altimeter, and compass on the left. Engine instrument grouping was good. The engine monitoring gauges (tachometer, EGT, oil pressure, and fuel totalizer) were located on the right lower half of the instrument panel. Readability and interpretation of these instruments were good.

The exploitation team found the warning lights poorly located and difficult to interpret. Landing gear warning lights were on the lower left sub-panel. The marker beacon, nose cone position indicator, "stabilizer ratio set for land" light, and trim warning placards were in the center warning panel. Fire warning and other lights were in the upper right portion of the instrument panel. Dimness of the warning lights, even at full intensity, caused interpretation difficulty. Color-coding was inconsistent throughout the warning/monitor indicators and red-colored warning light may or may not have been a normal condition. The monitoring and warning light system was adequate for providing vital information to the pilot.

Manual control of the nose cone, stabilizer ratio, and intake shutter doors provided pilot override capability for these normally automatic systems. Emergency airstart and landing gear controls were adequate, but required concentrated effort to actuate. The exploitation team incorporated an emergency hydraulic pumping unit for limited stabilizer control in the event of primary and boost pump failure. This system automatically actuated or manually by the pilot. Aileron control effected manually if the booster system failed.

Speed brake, gunsight electrical cage, and trim armament fire buttons were located on the control stick grip. Actuation of electrical cage, when pressing the trigger, was somewhat awkward; but did not necessarily limit the pilot's ability to operate the systems. The trigger was normally stowed in an upright position and was unfolded for operation. The brake handle arrangement was poor and of antiquated design,

Throttle controls rated good to fair. The positive lock lever for idle was good since inadvertent stopcocking of the engine was nearly impossible. The after burner engaging locking levers initially caused difficulty for the pilot because of the determined effort required to engage and disengage afterburner.

The MiG-21 appeared to have a high-speed ejection capability. The canopy designed semi-encapsulated the pilot during normal ejection sequence. Alternate controls allowed for separate jettison of the canopy. The ejection system was of unique design, partially encapsulating the pilot during the ejection. The seat formed an air blast shield to retain the front hinged canopy later jettisoned in the ejection sequence.

The MiG-21 ejection seats required several inputs such as pilot weight, height, etc., to dial in the proper barometric setting. By semi-encapsulating the pilot with the canopy during ejection, high-speed bailouts was possible without serious pilot injury. The system design operated at speeds up to 595 KIAS at sea level) and up to 2.05 IMN at altitude. From all indications, this ejection system was extremely effective and reliable.

Armor plating protected the MiG-21 pilot as indicated below:
- Headrest .68 inches thick
- Rear Plate .63 inches thick
- Front Plate .4 inches thick
- Glass Shield 2.5 inches thick

Review of all available combat gun camera film indicated that, although the MiG-21 tended to explode when hit by cannon/missile fire (probably due to wet wing design), the pilot ejected successfully in most cases. Effectiveness of this armor plating apparently contributes to the high pilot survivability rate.

The MiG-21 design incorporated a three wheel braking system. The nose wheel brake selected at the pilot's option. This increased the total system braking energy by 20 percent. After landing gear retraction, an automatic feature applied the wheel brakes to prevent tire rotation in the wheel wells.

During the exploitation team's verification of the MiG-21 gunsight radar capabilities, the gun mode obtained a 3-7 nautical miles maximum detection range in the missile mode and a 1.6 nautical miles maximum detection range.

The MiG-21 gunsight was ineffective during maneuvering flight. Manual ranging of the gunsight was not smooth or precise. System hysteresis and friction made it virtually impossible to prevent over control of the sight reticle diameter size with the throttle twist grip. Pipper jitter during cannon firing was in excess of 20 mils.

Gyro drift when tracking air targets was excessive. At g loads greater than +2.5, the sight reticle drifted to a point near the bottom of the sight combining glass. At very high g loads, the sight reticle disappeared entirely. The sight electrical cage functional was sluggish and slow to respond.

During air-to-air tracking, it was necessary to hold the electrical cage button (on the stick grip) until radar lock on occurred. The electrical cage button was poorly positioned and difficult to actuate when preparing to fire the cannon. Over-the-nose visibility restrictions limited the useful mil depression to 95 mils. Large lead angles during air-to-ground attacks with bombs, cannon, or rockets were not available. It was not possible to depress the gunsight in the cannon mode of operation as may be required for ground attack at long slant ranges.

Maintenance discrepancies

- 12 Feb 68   #1 Boost pump inoperative
- 24 Feb 68   EGT Malfunction (harness frayed)
- 28 Feb 68   Frayed brake cable
- 5 March 68   Oil System (6 sorties lost)
- 11 Mar 68   EGT Malfunction
- 27 March   EGT malfunction
- The oil system did not malfunction.[9]
- Still, only 11 sorties lost. The US jets did not come close to that.

---

[9] Unfamiliarity with the aircraft made a clogged oil filter seem like a major problem

Summary of technical evaluation of MiG-21 unique design features

- Very Low Wing Loading (50-55 psf)
- Lacquer Coating for Corrosion Prevention
- Ejection System (SK-1 seat and canopy)
- Air Intake (3-position, normal, Mach 1.5, Mach 1.9)
- Seat Position adjustable, but cramped
- Low Maintenance Requirements
- No roll, pitch, yaw stability augmentation
- Cooled Navigation Lights
- Optimized Cross Section
- Smooth surface finish only where it needed to be. Rivets protruded in out of sight locations.

Summary of tactical operational applications

Project flights conducted in an attempt to duplicate the Air Combat Maneuvering environment encountered in SE Factors contributing to test results that did not simulate this environment include:
- Participating pilots briefed on maneuver performance prior to each flight.
- Two-way UHF radio communication maintained between participating pilots.
- Engagements terminated when encountering unusual flight characteristics.
- Bingo fuel weights adhered to.

A life or death situation was not present; however, participating airplanes constantly flew to maximum performance. Only airplane performance was evaluated since missiles and guns not fired during engagements.

The exploitation results recommended:

The following general fighter tactics applied when engaging a MiG-21:
- Being aggressive. Using sound tactics while maneuvering for the advantage.
- Determining your opponent's ability—assuming that the "Red Baron" had been engaged until proven otherwise.
- Utilizing the Combat Spread Formation when flying a hostile area to provide visual coverage of each airplane's stern are engaged as a section to provide mutual support. The MiG-21 was extremely difficult to see due to its small size. The MiG-21 was normally under GCI control and positioned for a stern area attack.
- Maintaining a minimum of 450 KIAS while patrolling. This airspeed allowed instantaneous application of maximum g.
- Maintaining a high energy level while engaging. Trade airspeed for altitude only and never attempt a slow speed scissors. If necessary, dive away to regain airspeed for a reattack or to execute an escape maneuver.
- Forcing the fight to low altitudes to take advantage of the MiG-21's airspeed limitations and high stick forces below 16,000 feet.
- Using lag pursuit maneuvering close in. Because of the MiG-21's superior turning

performance, a close-in overshoot was highly probable if utilizing lead pursuit to close for a minimum range missile or gun shot. As the MiG-21 initiates a defensive hard or break turn, maneuver to a point 3,000 to 5,000 feet astern and outside the MiG-21's radius of turn. This prevents a close-in overshoot, reduces energy bleed-off, and places you in his blind cone. Continue maneuvering the stern area until it appears that the MiG-21 has lost visual contact then close for the kill.

• Maneuvering into the MiG-21's blind cone during all offensive maneuvering to capitalize on the MiG-21's visibility restrictions and to arrive in the aft hemisphere missile envelope.

• Avoiding dissipating energy by using hit and run attacks and Yo-Yoing high. Do not strive for a rapid close-in shot.

• Executing a high g roll away to position for a lag pursuit attack instead of performing a high Yo-Yo to counter the overshoot if a close-in overshoot is imminent during offensive maneuvering. This eliminates the possibility of a slow speed scissoring situation catching them.

• Using an oblique loop maneuver for reciprocal course changes below 16,000 feet once engaged; vice horizontal, high-Yo-Yo, or low Yo-Yo type turns. The oblique loop allowed the attacker to keep sight of the MiG-21 during the maneuver while capitalizing on the F-4/F-8's superior performance in the vertical plane. This maneuver repeatedly utilized was an effective position maneuver.

• When sighting a MiG-21, turn to engage head-on. Reduce lateral separation and jink to avoid cannon fire while closing. Turning to position the MiG-21 at a high TC if unable to engage head-on at that range, maintain this high TCA while accelerating for separation. Be prepared to break into an ATOLL missile or to negate a gun-firing pass and force an overshoot. If a wingman is present, he separates in the vertical to present two targets and employ loose deuce maneuvering in an attempt to sandwich the attacker.

• Reducing lateral separation between airplanes to a minimum if committed to a head-on attack. A MiG-21 can convert any lateral separation into a decreasing TC

• When passing the MiG-21 head-on, delay the turn back into him up to 5 sec or 90° of bogey turn, allowing the F-4/F-8 to accelerate and provide sufficient lateral separation again to meet the MiG-21 head-on after the reversal. A turn initiated immediately after passing losses energy and the MiG-21 gains TCA in the turn. As subsequent head-on high energy passes, continue the turning to dissipate the MiG-21's energy and make it vulnerable.

• When section maneuvering offensively to engage a MiG-21, close until the MiG-21 initiates a defensive maneuver. When the MiG-21 maneuvers, the wingman separates vertically. One airplane keeps the MiG-21 engaged, while the other employs loose deuce maneuvers in an-attempt to close for the kill.

• If the MiG-21 closed to gun tracking range (within 3,000 feet/30 degrees angle off), escape becomes difficult. Execute a nose low break into the attacker, accelerating to above 595 knots at maximum g for that speed; keeping the attacker in sight to effect an escape maneuver. If being fired upon, vary the g load and yaw on the airplane to negate tracking solutions. If the attacker follows you down into the high q low altitude region, a re-engagement may be considered. If the attacker rides high, do not pull back up to re-engage. Separate to evaluate the situation.

In addition to the tactics recommended above, the following tactics were pertinent to the

F-4 weapons system.
- A-visual retention of the MiG-21 beyond 2 miles was very difficult. AIM-7E/E-2 trigger squeeze minimum range thumb rules in the forward quarter were 3 miles and 2 miles respectively. Prior to firing an AIM-7 missile, radar lock-up required 4 seconds to allow radar settling and missile speedgate tuning. Resultant minimum ranges to allow for radar lock-up in the forward quarter were 4/3 miles for the AIM-7E/E-2 respectively. Since the probability of visual detection of the MiG-21 at 3-4 miles in the forward quarter aspect proved remote, successful simulated AIM-7 forward quarter firings resulting from an initial visual detection of the target did not occur. This did not include VID (visual identification) formation forward quarter simulated firings or CIC authorized firings.
- If the MiG was visible with 3-4 miles lateral separation, a successful simulated AIM-7E-2 shot was possible during the turn back if the lock-on accomplished prior to 45° to go. The time delay from radar lock-on with 45 degrees to go prior to roll out at 2 miles, head-on, approximates 4 seconds. This was normally sufficient time to launch an AIM-7E-2 missile head-on. The rapidly changing bogey azimuth and elevation in this aspect required a high level of crew proficiency.
- The F-4 two-man crew had a significant advantage over single seat fighters. Once engaged, the RIO was available to and had to concentrate on keeping sight of the attacker/attackers until radar acquisition was possible. The RIO repeatedly proved to be invaluable to the success of the engagements. Padlock lookout was mandatory against airplanes as small as a MiG-21.
- The energy advantage of the F-4 below 16,000 feet and above 450 KCAS allowed the F-4 to gain and maintaining the offensive if the techniques of lag pursuit, vertical reversal, and employment of slash attacks described in the General Fighter Tactics Section.
- If sighting the MiG-21 in the forward area closing, turn to meet him head-on with minimum lateral separation. If the MiG-21 did not turn to meet you head-on, assume he did not see you and turn for lateral offset in an attempt to convert to an aft hemisphere attack maintaining rigid lookout doctrine during the turn to Avoiding turning in front of a trailing wingman.
- If the MiG-21 turns toward you prior to passing abeam, turn into him to reduce lateral separation. Jink as necessary to negate a head-on gun attack. While closing, radar search the area behind the MiG-21. As the MiG-21 passes close abeam, drop a wing as necessary to keep the MiG-21 in sight. Maneuver in a maximum performance oblique loop to re-engage. Analyze the MiG-21 pilot's ability in this first turn. If he was not maneuvering aggressively, position yourself for a kill. If he maneuvered aggressively and has gained any advantage in the reversal, attempt again to meet him head-on. On passing, delay the turn back slightly, approximately 5 seconds (keeping the bogey in sight), to insure that sufficient lateral separation was available to compensate for turn radius and to insure a subsequent head-on pass. If the MiG-21 maneuvered and aggressively reverses on each pass, his energy level dissipates.
- If engaged above 16,000 feet force the fight to low altitude. Below this altitude, utilize the oblique loop to effect a turn reversal. The vertical reversal capitalizes on the F-4 energy advantage in the zoom and forces the MiG-21 to work the vertical and dissipate his energy. The F-4 speed on top was normally 250-300 KCAS. The MiG-21 does not regain his energy in the dive as rapidly as the F-4. When the MiG-21's energy level dissipates and/or you gain an advantage, continue pressing the attack but do not rush it. Capitalize on your performance

and exploit the MiG-21's limitations. Continue to perform slashing attacks; employing yo-yos until achieving a missile launch position. In the event a close-in over-shoot appeared imminent, execute a barrel roll or roll off maneuver to the MiG-21's blind area, approximately 1 mile aft in a lag pursuit attack. Retain your energy level. If a close-in overshoot develops, roll to effect separation as rapidly as possible, unload, keep the bogey in sight, and re-engage on your terms.

- When sighting a MiG-21 with separation available, meet him head-on, and proceed as above. If a head-on meeting was not possible, turn to keep the MiG-21 in sight, place him at a high TCA and accelerate. As the bogey closed, be prepared to break into a missile. Continue to keep the MiG-21 in sight, place him at a high TCA until a break turn was necessary to negate a gun attack and force a high angle overshoot. The MiG-21 will most probably Yo-Yoed high to conserve energy and maintaining the offensive.
- If the MiG-21 overshot flat and slid out ahead, a reversal or roll over was possible to arrive at his 6 o'clock position. Using extreme caution if not achieving immediate success. Be prepared to immediately unload into him and accelerate for separation. The MiG-21 was far superior at close in maneuvering than the F-4. Never attempt close in maneuvering unless it was apparent that the MiG-21 pilot was incompetent or holding a definite position advantage. The MiG-21 repeatedly demonstrates the ability to counter a high angle overshoot and rapidly regain a gun tracking position, if the U. S. fighter reverses into him as he overshot. Maneuver to his blind area, achieved a missile launch position, and immediately regain your energy level. Be sure the MiG-21 was not a decoy feinting poor performance. A competent MiG-21 pilot turned an apparently defensive situation for him into a very sudden offensive position. If a scissor situation was imminent, dive into his blind area and accelerate for separation.
- (KIAS) Keep the MiG-21 in sight. Obtain sufficient lateral separation to reverse back into the MiG-21 to effect a head-on engagement. Attempt to acquire the MiG-21 on radar in the turn for a possible Sparrow shot. The RIO had to remain padlocked on the MiG-21 until it was well within radar gimbal limits (45 degrees and the pilot had confirmed visual contact). When the RIO goes to the radarscope, the pilot padlocked on the MiG-21 and coached the RIO on his position. As the MiG-21 approaches the nose, PLM may employ if the RIO was unsuccessful in acquiring a radar lock-on. Feel out the MiG-21 during subsequent maneuvering and effect a kill when and if presented the opportunity.
- The exploitation established three rules as essential to successful F-4 section tactics against the MiG-21 or any small, low wing loaded airplane. They are:
- All crewmen had to maintain visual contact with the bogey. The RIO could not return to the scope until the bogey was within approximately + 45 degrees of the nose and the pilot had confirmed visual contact (2) Engage only in section. It was very easy to split the F-4 section, lose mutual support, and fight two separate one-on-one engagements. In combat, when the possibility of multiple bogies existed, splitting the section could be disastrous.
- A steady flow of information had to take place within the F-4 section. This required relaying relative position to each other, relative position of the bogey, intentions, and tactical orders.
- Mutual support between F-4s engaging a MiG-21 dictated that each airplane in the section be able to protect and support the other during an attack. Each member of the section had to have sufficient but not excessive separation to launch missiles at any threat posed. In addition, the F-4 section required positioning to prevent the bogey from working both F-4s

as a unit, while maintaining contact with each other.
  • Apply maximum separation between F-4s-3 miles on VID formation, maintain maximum separation between F-4s abeam, co-heading-1 mile, and offset heading on the offensive-about 90 degrees. This sets up a two-on-one offensive attack that forces the bogey to meet threats from divergent angles.
  • After a head-on pass, the F-4 section had to maneuver as necessary to maintain mutual support and visual contact.

Exhaust smoke comparison

AFFTC lessons learned

- Power checked at Mil power prior to brake release. Brakes failed to hold in afterburner
- Rudder effectiveness occurred at 45 knots
- Nose wheel liftoff at 114 KIAS (with full aft stick)
- At 15,400 pounds, with 30 degree (full) flaps, takeoff speed was 165 KIAS
- Afterburner failed to light when selected until after military thrust was achieved
- Stabilator was the only trimmable control surface
- The engine never stalled

ADC comments on Project HAVE DOUGHNUT

ADC found visual contact difficult because of small size, except the silver color made the planform view relatively easy to see. The radar signature to MA-1 fire control system indicated contact and tracking adequate for Intercept completion with contact 20-25 miles in all aspects. ADC found stern contacts were best and head-on co-altitude the worst. Relative to an F-4, it provided a return 1/2 as large in front, 3/4 in beam and almost identical in stern because of engine modulation. The IR signature to MA-1 fire control system was adequate for acquisition and tracking. ADC found it about 3/4 as strong a return as the F-4 and similar to the F-106 in military, but not as strong as F-106 in A/B afterburner. ADC found the performance of the test aircraft not as good as expected with limiting factors capable of exploitation being slow engine response in military, me required for A/B initiation, q limit below 15,000', visibility to rear, visibility over the nose. Another area needing exploiting included energy and time required to transit the transonic zone with pylons and missiles aboard, stiffening of controls at low speed and low altitude, endurance at maximum power, and one radar fire control system.

ADC found the F-106 aircraft's radar capable of acquisition and usable to put the F-106 in position for armament launch. The F-106 used radar snap-up attacks with all-aspect armament load to exploit test the MiG-21's lack of adequate fire control system and the inability of the MiG pilot to see over the nose and through the windscreen. The ADC recommended using the F-106's ability to accelerate faster than the MiG-21 to achieve a higher speed (beyond q limit) to separate anytime it found the F-106 not in an advantageous position during engagements. The F-106 could use missile launch and then use lag pursuit while closing to gun kill position, depending upon its superior turn capability to pull necessary lead for gun firing. ADC recommended expediting procurement of cannon for F-106 for near term close-in-kill armament, and use the superior zoom capability of the F-106 to advantage for repositioning after separation during engagements.

The ADC cautioned the bar overhead in the F-106 canopy causing F-106 aircrews to lose sight of the MiG-21 during close-in engagements. With this defect, roiling maneuvers could get the F-106 into trouble and every effort made to expedite replacement of the F-106 canopy bar with a clear pane.

Following the exploitation, ADC cautioned F-106 aircrews to take care to preclude unnecessary expenditure of energy when observing the MiG-21 initiating a turn as the appearance of generation of a great amount of turn was deceiving. The misleading size of the MiG-21 caused error in estimation of range and rate of closure by F-106 aircrews, thus the F-106 should not attempt slow speed turning contests with the MiG-21 as performance was close to equal and a slight miscalculation could be fatal. The F-106 should keep its speed at 400 to 450 KCAS during patrol and during an engagement.

The exploitation determined that the F-106 could use its missile armament during an engagement with efforts to modify the fire control system with IE "boresight expedited. The IR "boresight modification should include automatic radar look-on by caging the radar antenna to the IR head and sweeping the range gate out effecting radar look-on with the option of caging the radar antenna dead ahead.

The ADC exploitation of the MiG-21 found the F-106 patrol formation, as taught at the Interceptor Weapons School, provided adequate protection against surprise attack by the test aircraft. It recommended improving F-106 radar and all flight members know visual search patterns to insure responsibilities for sector search. The evaluation determined radar tracking procedures lacking and the need for procedures for assigning responsibility within and between elements for armament launch when some or all members of the flight acquire a MiG-21. The F-106 aircrews needed training that when pressing an attack after acquiring the test aircraft, they had to expend armament in order of priority, i.e., missiles, then press to gun position or separate if no gun is aboard or all missiles are expended.

The ADC found an all-aspect armament capability with successful qualification of 5 out of 5 missile simulator evaluators (WSEMS) during the test, which indicated the AIM-4F missiles properly prepared will see the test aircraft as far out as 3.5 miles on front aspects with a high probability of successful guidance. This indicated the F-106 with its present configuration and with developing tactics being an effective counter to the MiG-21.

Findings: the bottom line

Air Force Systems Command (AFSC) personnel evaluated the performance and handling qualities of the MiG-21. The Air Force and Navy used photos such this for briefings with the HAVE DOUGHNUT pilots at Groom Lake.

- Simplicity; Ease of Flying–It's a good, honest aircraft
- Reliability and Maintainability (20 minute turn around)
- Cross-Sectional Area
- Engine exhaust smoke was another exploitable discovery. The FISHBED's after burner marked the aircraft's location by producing white puffs of unburned fuel when it engaged or disengaged. This was a small consultation because the American jet fighters similarly aided the enemy by leaving a smoke trail that exposed their presence.
- 3-wheel brake concept.
- Armament was adequate for an interceptor. For the point interceptor role, the MiG-21's basic weapons included a 30 mm cannon loaded with 60 rounds of ammunition and 2 AA-2 ATOLL heat-seeking missiles. The Soviet-built ATOLL missiles were copies of the US-made AIM-9 Sidewinder obtained and reverse engineered by the USSR when one lodged in a MiG-17F and failed to explode.
- In an air-to-ground role, the MiG-21 carried the 30 mm cannon and two pods containing 32 57 mm folding-fin aerial rockets. The cannon proved potentially lethal against tanks,

however encountered considerable piper (gunsight) jitter while strafing. Another drawback for the MiG-21 was its high speed-low altitude stability in rough air.

- Excessive bleed-off during high-g turns turned into an advantage for the MiG by decreasing its turn radius while sustaining its g force at slower speeds than comparable US fighters. Obviously, in a turning fight, this gave the FISHBED a tactical advantage.
- HAVE DOUGHNUT confirmed that in actual combat in the skies over North Vietnam that the American fighters flew faster than the MiG, and that the MiG pilots overcame this US advantage by tactfully drawing the US fighters into turning engagements where the superior speed did not matter. The MiG knew the appropriate angle to cut their circle and make the MiG guns effective. Maintaining high speed at low altitude was the key to MiG survival.
- The Navy created the TOP GUN Weapons School in 1969 and experienced strong results against the MiG-21 when they encountered it in 1972.
- The Air Force did not create a dissimilar air combat program until 1972/73 when the Air Force created Red Flag to give its pilots a better edge in the fight. The USAF finished the war in 1973 with a two-to-one kill-loss ratio, downing 137 MiG while losing 65 aircraft, including bombers, to MiGs.
- G-load factor was 8 g without stores, 6 with stores.
- Max indicated airspeed-595 knots below 15,000 feet 640 knots above 15,000 feet;
- Maximum indicated Mach _ 2.05 without stores, 1.6 with stores.
- Strike radius-370 nautical miles with external fuel.
- Poor forward and rearward visibility   F-4 acquired at 3-5 miles range.
- Low Q limit-Below 15,000 limited to .98 Mach or 595 knots–severe buffet.
- Afterburner puff-Above 15,000 FISHBED E produced a puff in/out of AB
- Engine response-extremely slow.
- Cockpit noise-extremely low.
- Gunsight capabilities-3.7 Nautical Miles, missile mode: 1.6 Nautical Miles, gun. Gun/missile target tracking impossible over 3 g.
- Slow speed   -The MiG-21 could maneuver at 115 KIAS.
- The MiG-21 engine was technologically behind its US counterparts, so spool-up from idle to full military power required 14 seconds, with a tendency to hang up in the process. This often led to hot compressor stall or engine over-temperature conditions.
- Low Altitude Transonic Vibration. Typical of delta wing aircraft, excessive high g turns caused airspeed bleed-off.
- Formation flying proved to be difficult because of the MiG's slow engine spool-up handicap. It took 14-seconds to accelerate from idle to full power and formation maneuvers required constant use of speed brakes and rapid throttle movement.
- Flying in Turbulence caused poor directional stability.
- Night Flying.
- Gunsight was deficient. The tracking index drifted off the bottom of the windscreen when tracking targets in excess of three G.
- Exceptionally heavy pitch force required above 685 mph or a .98 indicated Mach Number.
- Problematic recovery during dive-bombing, strafing, or air-to-ground rocket firing.
- Exceptionally slow engine acceleration from idle to full military power.
- Design to prevent overstress problems during pull up from a target made it difficult or

impossible to achieve high pitch rates expected of a fighter-bomber. Heavy pitch forces at high speed limited the pilot's ability to recover from a diving attack or maneuver while approaching and departing the target.

• Most significantly was the aircraft not being able to go supersonic below 15,000 feet. Severe buffeting at low level prevented it from exceeding 685 mph or .98 Mach airspeeds. This airspeed limitation exposed a major exploitable design flaw that F-4s and F-105s exploited during the Vietnam War by typically approaching the MiG-21s at 633 mph then departing supersonic.

• Easy to kill -non-sealing tanks, unprotected engine, light metal structure, high pressure O2 bottles–85% kill probability.

• A special limitation for the day-visual was the fighter-interceptor's front and rear visibility. The combination of a bulletproof glass slab and the windshield restricted forward visibility through the gunsight. The protective seat flap over the pilot's head and the narrow design of the ship's canopy and fuselage structure handicapped visibility in the 50-degree tailcone.

What the planes should do if encountering a MiG-21

• If you are flying an F4C, D, or E and a MiG 21 finds you below an altitude of 15,000 feet you had better run away. Speed is life.

• If you are flying an F-105 and a MiG 21 finds you, avoid prolonged maneuvering engagements. Get into the MiG-21 blind cone, use mutual flight support, and use hit-and-run tactics. Maintain maximum airspeed below 15,000 feet and do not get in a turning fight. Speed is life, and you can only go in a straight line.

• If you are flying an F-111 and a MiG 21 finds you, remember that the MiG has superior turn performance everywhere. Below 15,000 feet, the MiG-21 has superior level acceleration up to its speed limit. When attacked get below 15,000 feet and above the MiG-21 speed limit of .98 Mach. Again, speed is life, and you can only go in a straight line.

• If you are flying an F-100D and a MiG 21 finds you, remember that the MiG-21 has significant turn performance and level acceleration advantages. Avoid maneuvering, use hit-and-run, mutual support, and again, Speed is life, and you can only go in a straight line.

• If you are flying an F-104 and a MiG 21 finds you, remember that the MiG-21 has superior turn capability. Avoid maneuvering, use hit-and-run, mutual support, and defend yourself by accelerating above MiG-21 limit speed. Speed is life, and you can only go in a straight line.

• If you are flying an F-5A, you can closely simulate the MiG-21 up to mach 1.2 for combat engagements.

• If you are flying an RF-101 and a MiG 21 finds you, remember that the MiG-21 has a slight advantage in afterburner and superior turn capability everywhere. When caught by a MiG-21 dive and run, speed is life.

• If you are flying a B-66 and a MiG 21 finds you, you are a target.

• If you are flying an RF-4C and a MiG 21 finds you, dive and run, speed was life

• If you are flying a Navy F-8E and a MiG 21 finds you, the MiG can out turn the F-8 in a close in fight. Speed if life.

• If you are flying a Navy A-4F, A-6A, or A-7A and a MiG 21 finds you, the MiG-21 pilot can chose to engage or disengage you at will. You are a target, but with a chance to live.

CHAPTER 4- MiG-17F Projects HAVE DRILL/HAVE FERRY

The HAVE DRILL MiG-17F designated YF-113A

The HAVE FERRY MiG-17F designated YF-114C. The Project HAVE DRILL and HAVE FERRY MiGs arrived at Groom Lake on Mardh 12, 1969. The two aircraft were obtained from Israel after two Syrian pilots mistakenly landed on an Israeli airstrip. Lt. Col. Wendell H. Shawler, chief of the 6512th Test Squadron's Special Projects Branch at Edwards Air Force Base, served as project manager. Participating pilots were VX-4 test pilots Lt. Cmdr. Foster S. "Tooter" Teague and Lt. Cmdr. Ronald E. "Mugs" McKeown, and Maj. Fred Cuthill

Background of the MiG-17F technical exploitation

Following the successful exploitation of the HAVE DOUGHNUT MiG-21 FISHBED a year earlier, considerable interest focused on an even more intense exploitation of the older MiG, the MiG-17F widely exported and operationally deployed to far eastern nations within the Communist sphere. In the Vietnam War, the cannons of the MiG-17F continued to dominate the air war against the more superior aerial assets of the US. This limited war environment made the exploitation of operational Communist systems a vital necessity. The FTD at Wright Patterson recognized its basic mission including the evaluation of foreign material of scientific and technical value to both the national intelligence community and Air Force research and development organizations.

US aerial combat losses over Southeast Asia further stressed the MiG-17F threat and demonstrated the need of a tactical as well as technical evaluation of this weapon system. Complete volumes of handbooks and specifications were available in the intelligence community on the MiG-17, but the exploitation team needed to evaluate the actual aircraft to obtain vital tactical and operational information necessary for the effectiveness and survival of their air warfare exploitation teams. At the same time, exploiting an operational plane would provide the opportunity to re-evaluate current hand-book-data, give their technical personnel an excellent insight to past levels of Soviet technology, and allow their more clearly predicting future advances. Likewise, the exploitation of current threat aircraft such as the MiG-17F was extremely important because of the research and development application it would have on their own technological effort in future developing and testing counter-threat systems.

The HAVE DOUGHNUT Special Projects veterans based at Area 51 recognized the need for further exploiting the aerial assets of their enemy and honed their individual skills and equipment to provide this service to their customers. Customers for this project were Navy, USAF, and Intelligence customers, i.e., the FTD of AFSC (a.k.a. National Air & Space Intelligence Center (NASIC)), AFFTC, Air Defense Command (ADC),, SAC, Aeronautical Systems Division (ASD), Tactical Air Command (TAC), and Navy Weapons Center (NWC). The mutual need for acquiring a MiG-17F ended four months after HAVE DOUGHNUT when two Syrian MiG-17Fs lost on a navigation exercise inadvertently landed at an Israeli air base on August 12, 1968, where flying two MiG-17F FRESCOs, disoriented Syrian first lieutenants Walid Adham and Radfan landed in Israel by mistake.

This acquisition of the MiG-17s was of high importance to the United States as the nimble MiG-17, though slow and dated, represented the main adversary type encountered in the skies of Vietnam. Though limited to subsonic performance, VNAF MiG-17s flying circles around American fighter pilots lead to dismal kill ratios. After testing, the Israelis turned the two MiG-17s over to the United States for analysis. In 1969, the exploitation team already assembled at Area 51 added the two ex-Iraqi MiG-17s transferred from Israeli stocks to the operation under project names HAVE DRILL and HAVE FERRY using US designations and fake serial numbers to identify them in DOD standard flight logs.

The Project HAVE DRILL code name identified a joint Air Force/Navy program to evaluate and exploit the capabilities of MiG-17F aircraft and its power plant, the VK-1F engine, both developed in the early 1950s. Initially produced by other Communist nations. MiG-17F as a "day fighter," the MiG-17F was still used extensively.

The FTD of the Air Force managed the overall project with the Naval Scientific Technical Intelligence Center coordinating the Navy's effort.

FTD maintenance performed on the MiG-17F HAVE DRILL and HAD FERRY MiG-17s began with the preflight and technical phase of Project HAVE DRILL.

The HAVE DRILL aircraft arrived at the Area 51 exploitation site on 27 January 1969, where uncrating and assembly began on 29 January 1969. During assembly, FTD examined the aircraft for discrepancies and after assembly, performed a 50-hour inspection. FTD then repaired, adjusted, and operationally checked during the inspection the noted aircraft discrepancies to render the MiG-17F ready for flight in 19 days.

NOSE AIR INLET DUCT, SPLIT

Discrepancies found during technical exploitation.

- Right hand brake air line damaged beyond repair,
- Right hand main gear door actuator hydraulic hoses deteriorated.
- Left hand main gear door actuator hydraulic hoses deteriorated,
- Left horizontal elevator damaged at trailing edge,
- SRD-IM Radar inoperative.
- SIRENA coax cable to antenna lead broken.
- SIRSNA-2 unit #3 cannon plug loose and not safetied.
- APT fuel tank transfer fuel line leaking aft of quick disconnect.
- APT fuel tank vent line leaking APT of quick disconnect.
- Oxygen line in nose hay leaking,
- HF inverter removed for troubleshooting and adjustment,
- Seat ejection cartridge outdated,
- Air line on gun package damaged.
- Speed brake selector valve leaking.
- Hydraulic leak in left wing at fuselage.

During the assembly and 50-hour inspection, FTD personnel inspected, adjusted, repaired, and operationally checked all the systems while installing or altering the instrumentation necessary to exploit the technical aspects of the plane.

Description of the MiG-17 test item

There were presently four operational variants of the FRESCO; however, the exploitation team confined this description to the FRESCO C version utilized in the conduct of this test. The following physical dimensions apply to that aircraft.
- Wing span 31.5 feet.
- Length 38.32 feet.
- Height 12.47 feet.23 mm
- Diameter of fuselage 4.76 feet maximum.
- Take-off gross weight (clean) 11,803 pounds.
- Take-off gross weight with external fuel: 13,414 pounds.

Sweep of the wings varied from 49 degrees inboard to 45 degrees outboard. The horizontal tail swept back 45 degrees, while the vertical portion swept 55 degrees, 41 minutes.

The VK-1F engine powered the FRESCO C, an afterburning version of the earlier VK-1. The engine was a centrifugal flow compressor type, with a two-position nozzle positioned full open for maximum power operation, and full closed for military power. The thrust rating varied from 5,720 pounds at military power to 7,440 pounds at maximum power.

One 37 mm internal cannon and two 23 mm internal cannons, all nose-mounted, comprised the primary armament of the FRESCO. The 37 mm cannon had a capacity of 40 rounds, while the two 23 mm cannons hold 80 rounds each. Total continuous firing time available with a full load of ammunition approximated six-seconds. An optical gyroscopic lead computing sight, utilizing a range only radar, provided the pilot with a solution to the firing problem. The exploitation team knew from observation that FRESCO C carried two externally mounted general-purpose bombs instead of external fuel tanks when serving in a tactical role.

Simple, but adequate, avionics characterized the FRESCO C. The range only radar defined required range information for the pilot The FRESCO had no radar on board. It had the common X-band passive tail warning radar receiver, designated Sirena with a nominal range of five nautical miles when illuminated by a peak power source. The inclusion of an SHO-2 L-band transponder permitted an identification Friend or Foe (IFF) capability.

The purpose of this test was to determine the best USAF tactical aircraft offensive and defensive maneuvers against the FRESCO C (MiG-17F) through flight test to verify/modify data and conclusions contained in tactical doctrine.

The MiG-17F pilots

The MiG pilots were USAF Col Robert `Bobby' Bond (who later flew A-7Es in Southeast Asia), Navy Commander Marland W 'Doc' Townsend (an F-4 Phantom II pilot), Navy Commander Tom Cassidy (a future admiral), and Navy Commander Foster S 'Tooter' Teague (who later commanded a carrier air wing in Vietnam). The two MiG-17s flew 198 sorties (usually together) against a variety of US Navy warplanes, ranging from the F-8J Crusader to the A-6A Intruder. Later, separately, they flew against the USAF F-102A Delta Dagger, F-104A Starfighter, and F-106A Delta Dart.

The shock from exploring the MiG-17

The HAVE DRILL and HAVE FERRY evaluation details of testing two MiG-17s were minuscule, causing some to question the reason for testing this older vintage aircraft from the Korean War era, especially since volumes of handbooks and specifications were available on the MiG-17F in the intelligence community. This line of thought changed somewhat when pilots found how the canopy-mounted periscope in the MiG-17F improved visibility to the rear hemisphere. This and other hands-on observations convinced the Air Force of evaluation of the actual aircraft furnishing vital tactical and operational information necessary for the effectiveness and survival of their air warfare exploitation teams. The shock and wisdom of this decision to go with the program came to fruit where the MiG-17F scored kills of 100% against Navy fighter pilots on their first MiG-17 challenge during the exploitation at Groom Lake. On their first encounter, Air Force did not score much better.[10]

A year earlier, both the USAF and USN had learned through the HAVE DOUGHNUT exploitation to avoid prolonged maneuvering engagements, aka dogfighting. HAVE DOUGHNUT encouraged emulating the MiG-21's hit-and-run tactics. The MiG-21 encounters over the skies of Nevada had developed dozens of ways in which the F-4E Phantom, flown against the MiG-17F in 26 sorties, and the F-105D and F-105E Thunderchief, flown 18 sorties, were superior to the Soviet fighter. Despite its sleek shape, the F-4, F-105D, and F-104 found the MiG-21's performance at high altitude inferior. Exploitation of the MiG-21, which fought with missiles, did not solve the problem of the US continuing to lose the air war in Vietnam against the older MiG-17F that shot only cannons.

---

[10] These embarrassing encountered resulted in the Navy creating the TOP GUN Weapons School in 1969, which produced strong results against the MiG-21 when they encountered it in 1972. The Air Force followed the Navy's lead by creating a dissimilar air combat program in 1972/73 when it created Red Flag to give its pilots a better edge in the fight.

Technical exploitation of the MiG-17F FRESCO

As part of the technical phase of the exploitation, AFFTC personnel installed the instrumentation to obtain quantitative performance and stability data of the HAVE DRILL MiG-17. In addition to the airborne data acquisition system, they installed various temperature and pressure probes to measure engine inlet conditions during engine ground runs. Installation of the test equipment in the MiG-17F began on 28 January 1969, and completed on 16 February 1969; with approximately 960 manhours required for the installation. This figure did not reflect the in-shop time expended by the AFFTC calibrating test instrumentation components, fabricating special adaptors, fittings, fuel lines, etc., which could not be accomplished on-site due to the lack of the specialized equipment required to accomplish these tasks.

Installation and/or alteration

- A UHF antenna, wiring, mounting, brackets, and radio set installed.
- A tape recorder installed in the cockpit.
- Cockpit installations and modifications included the "G" meter on left hand aux. panel.
- Intervalometer on right hand aux. panel.
- Airspeed Indicator-Main panel.
- Altimeter-Main panel.
- Machmeter in place of radio compassed indicator.
- Removed ships EGT and Tach indicator, calibrated and reinstalled.
- Removed V.H.P. control picked up plugs and wiring to go through the cockpit pressure area to nose section for power leads.

• Installed instrumentation circuit protector panel,

Nose section installations and modifications include the following:
• Removed VHF radio and installed a photo panel.
• Pitot-static system mounted on nose and spliced in at wing disconnect.
• Outside air temperature probe mounted on oxygen servicing door.
• X-band antenna mounted on oxygen servicing door.
• Gun platform installations and modifications include the following:
• Ammunition cans removed.
• Manufactured brackets for and installed a Midwestern oscillograph model #581 fourteen channel.
• Signal conditioner package mounted near oscillograph.
• A K-3 Minneapolis Honeywell attitude gyro installed.
• Three rate gyros installed.
• Two potter flowmeter amplifiers and totalizers installed.
• Oscillograph counter timer delay box installed.
• Fuel amplifier for oscillograph mode installed.
• X-band beacon and power divider installed.
• X-band antenna installed.
• Accelerometer transducer installed.

Test equipment set up in Area 51 hangar

Engine Compartment Installations and modification include the following.
• Flowmeter transducer installed in main line.
• Flowmeter transducer installed in afterburner line.
• Wiring tapped into ships tach generator.

- Control Surface Position Installations.
- Markite potentiometers and brackets mounted at elevator torque arm,
- Markite potentiometers and bracket mounted on rudder torque arm.
- Markite potentiometers and bracket mounted on left hand aileron torque arm.

Technical exploitation objectives

The objective of their joint exploitation efforts focused on data acquisition internal to the plane and by the Special Projects exploitation team tracking the planes during the tactical phase to follow. The data collection included manual, voice recording, photo recording, and electronic data via an oscillograph. In the MiG hangar, AFFTC installed an external observed r's panel to record engine ground run data in addition to photographing everything by a hand-held 35 mm camera.

AFFTC installed manual system-calibrated test instruments to replace the standard aircraft instruments in the pilot's panel. These instruments included:
- Clock: 8-day with start and stop sweep second hand (Type A-13).
- Indicator, Airspeed: 50 to 850 knots, Kollsman Instrument Corp. (Type 739DX).
- Altimeter: 0-80,000 feet, Kollsman Instrument Corp., Type AN5760-5.
- Machmeter: 0.5 to 1.5 Mach No., Kollsman Instrument Corp., Type A-2.
- Accelerometer:-4 to 10g, Bendix Aviation Corp., Type B-6.

In addition to the above, AFFTC calibrated the standard aircraft engine RPM and exhaust gas temperature indicators, calibrating the EGT as a system utilizing a Jet cal tester. To obtain the proper heater probes, the AFFTC personnel sent a sample thermocouple probe to AFFTC at Edwards to check against various probes on hand.

To record pilot comments, the exploitation team installed a Wollensak Model 4200, cartridge type magnetic tape recorder on the aft end of the right-hand vertical console. AFFTC mounted this recorder on a quick-release mount to facilitate changing tape cartridges and to provide access to the various ground servicing valves. Input to the recorder was through a special adaptor installed between the pilot's helmet disconnect and the aircraft's disconnect cable. Use of this adaptor allowed the pilot to start and stop the recorder as required and enabled recording of all transmissions and receptions when the recorder was running.

In addition to the Wollensak, Model 4200 recorder installed in the aircraft, AFFTC test engineers utilized another Model 4200 in the Control Tower as a backup measure to record all transmissions and receptions during USAF test flights.

To record basic performance data, AFFTC installed a 12-parameter folded photo recorder laterally in the nose bay between fuselage frames 3 and 4 in the space made available by the removal of the aircraft's standard ADF, VHF, and IFF radio transceivers.

AFFTC used a Traid Corp., Model 1000B 16 mm pulse-recording camera equipped with an Angenieux, 10 mm, fl.8 lens and an LB4BT, 50-foot capacity film magazine, to photograph this panel, and Eastman Kodak Plus X, l6mm film as the recording media.

Twenty-two GE 1495 type 28 VDC bulbs mounted on the face of the panel provided indirect lighting of the test instruments. They painted the top 1/3 of each bulb with high temperature silver paint to prevent glare into the camera lens and to act as a reflector.

The pilot selected a USAF B9A intervalometer mounted at the top right hand side of the pilot's instrument panel as required to pulse the photo recorder camera at a pulse rate selectable from two frames per second to one frame every 60 seconds.

AFFTC personnel mounted an oscillograph and associated signal-conditioning equipment on the gun pod between aircraft frames 4 and 9 by removing the ammunition containers and installing special provisions for mounting the test equipment. An oscillograph run switch mounted on the test instrumentation control panel controlled the oscillograph. In addition to this switch, AFFTC mounted an OSC FAIL indicator light adjacent to the OSC RUN switch. This light illuminated to indicate an oscillograph malfunction or exhaustion of the oscillograph paper. The oscillograph system recording the stability parameters while a Midwest Instruments, Inc., Model 581, l4-channel oscillograph recorded the fuel flow data.

AFFTC measured engine and afterburner fuel flowed using Potter Aeronautical Corp. turbine type fuel flow transmitters, Model 130 converters, and a special AFFTC designed oscillograph signal conditioning module. Modified Model 130 converters converted the cyclic output of the fuel flow transmitters to suitable outputs for oscillograph and photo recorder presentation. Sodeco 5-digit counters presented fuel used data on the photo recorder with each count representing approximately 1/10 gallons of fuel used, presenting the fuel flow rate data on the oscillograph as a cyclic function. To aid in data reduction, a pedestal imposed on the oscillograph every 64 cycles, with each 64 cycles representing approximately 1/10 gallons of fuel used. Using the oscillograph time lines, AFFTC determined fuel flow by referring to the fuel flow calibration chart plotted as CPS against flow rate.

AFFTC also fabricated new fuel lines incorporating Potter flowmeters prevented modification of the aircraft's standard fuel lines, and enabled adapting from metric size to standard AN size tubing and fittings. A one-inch diameter transmitter installed in the main fuel supply line between the tank outlet and the fuel filter inlet measured total fuel supplied to the engine and afterburner. They measured afterburner fuel flow using a 3/4 inch diameter transmitter mounted in the afterburner fuel line between the afterburner fuel control unit and the afterburner spray ring inlet, and obtained the afterburner operating, engine fuel flow by subtracted afterburner fuel flow from total fuel flow. Fuel evaluations included fuel temperature and outside air temperature.

To evaluate engine RPM, AFFTC removed the standard aircraft RPM Indicator and tachometer generator from the aircraft and calibrated them as a system at the AFFTC facility at Edwards AFB in addition to calibrating the standard aircraft tachometer indicator,

AFFTC displayed the parameters of indicated airspeed and altitude in the photo recorder on Kollsman IAS and altimeter indicators. These instruments connected thru the used of flexible surgical tubing and tee fittings to the aircraft's normal wing boom total and static pressure systems.

To record control surface positions, AFFTC recorded elevator, rudder and left hand aileron positions recorded on the oscillograph, measuring them using Marklte 5000 ohm, 357 degree travel potentiometers as transducers. They measured rudder and elevator positions using a cable and pulley system tied directly to the control surface torque rod, and the left hand aileron position using a cable and pulley system tied to the control linkage just inboard of the aileron.

AFFTC also used the oscillograph to record aircraft bank and pitch angles, measuring them using a Honeywell K-3 cageable attitude gyro, type JF7044A-35 mounted on the gun pod. A gyro cage-uncage light and switch mounted on the test instrumentation control panel provided data on the gyro limits.

The oscillograph recorded roll, pitch and yaw rates as measured using Humphrey rate gyros. The oscillograph recorded aircraft normal acceleration recorded using a Statham A43-10-350 unbonded strain gage transducer mounted on the aft bulkhead of the gun bay, marking an event condition with a special push button type switch mounted on the front left hand top side of the pilot's control stick. Actuation of the switch deflected a trace on the oscillograph and illuminated a light in the photo recorder.

AFFTC used a 14 channel, modular type signal conditioning package for conditioning of the various transducer outputs to acceptable levels for oscillograph recording and to provide automatic in-flight "R" cal capabilities. AFFTC used a signal conditioner designed and built by for use in the Test Pilots School stability seat packages.

AFFTC conducted four engine runs to obtain quantitative, temperature and pressure information of various engine parameters, and to determine installed engine thrust. Two engine ground runs obtained engine thrust and static pressure, temperature information-one to obtain dynamic measurements of pressures and temperatures for the Navy Propulsion Lab and one to obtain fuel flow data at various power settings. Necessitating the last run was the failure of the oscillograph to record during the engine thrust run, with resultant loss of fuel flow data.

The exploitation team used two different systems of tie-down to secure the plane. The first method used two, approximately 20-foot long 3/8" diameter, steel cables attached to the main gear up-lock fittings on one end with the other ends shackled to another cable approximately 20-feet long running to the tie-down attachment point. This system resulted in the cables attached to the main gear up-lock fittings riding on the main gear tires, placing the strain link used to measure thrust directly in the exhaust blast. To eliminate these problems, AFFTC fabricated new cables and fittings to secure the plane for thrust measurements using AFFTC designed and fabricated 15,000 lb. strain links inserted in the tie-down line. They connected Mayberry Instruments, Inc., multiple turn self-balancing indicators, calibrated in pounds of thrust (0-7,000 lbs) to the strain links thru a 30-foot electrical lead. They mounted the indicators at the external observed r's station and manually recorded the readings. This method appeared to be a repeatable and accurate method of obtaining thrust measurement in the field. To reduce hysteresis error at each stabilized point, the testing personnel physically rocked the aircraft forward and aft at the left hand wing tip.

The Naval Air Propulsion Test Center required the measuring of various engine parameters during throttle bursts for which the exploitation team conducted one engine ground run. NAPTC paralleled its test recording equipment into existing test instrumentation systems to measure compressor inlet pressure, engine RPM, and fuel flows. The Navy installed an X-Y plotter and a CSC direct-write oscillograph in a radio-equipped truck to enable communication with the aircraft operator during the engine run. To conserve engine ground run time, the Navy combined this run with the thrust measurement run, using FTD personnel to manually and photographically record the tests using an open face test instrumentation panel mounted in a radio-equipped vehicle and interconnected to the aircraft through 40-foot long pressure tubes and thermocouple leads.

In addition to the pressures and temperatures measured on the engine, FTD personnel fabricated a pressure, temperature rake containing three pressure probes and one temperature probe to measure exhaust gas pressures and temperature at the exhaust nozzle exit. Mounting the rake on a type B2 bomb hoist enabled removal of the probe from the gas stream during afterburner operation. Outputs of this rake connected to the test engineer's panel through 40-foot pressure and thermocouple leads.

The Aeronautical Systems Division and Research & Technology Division conducted the systems evaluation, vulnerability study of the MiG-17F as an exploitation team effort. The effort included the combined results of the Aeronautical Systems Division (ASD), Air Force

Astronautics Laboratory (AFAL), Air Force Flight Dynamics Laboratory (AFFDL), Air Force Material Command (AFML), Air Force Aero Propulsion Laboratory (AFAPL) exploitation conducted during project HAVE DRILL. ASD Laboratory objectives under this program were to determine technological state-of-the-art of the aircraft and its subsystems, determine the vulnerability of the aircraft to the US current gun and missile threat, and determine characteristics of the installed radar's performance and its susceptibility to ECM.

Subsystem analysis of the hydraulic and Pneumatic systems of the FRESCO C (MiG-17F) hydraulic system consisted of a main system and a booster system. The main system supplied fluid power for landing gears, flaps, dive brakes, and afterburner nozzles. An engine-driven, gear-type, fixed delivery hydraulic pump rated at 5-3 g, Pi at 2500 rpm powered the system. An engine bleed air pressurized reservoir supplied fluid to the pump at pressures varying between 1137 and 1190 psi. Fluid from the pump flowed through two felt type filter assemblies, mounted in parallel, to the automatic relief valve, then to the pressure accumulator mounted on the right side of the fuselage at the wing root, and to the hydraulic actuating cylinders. When not required the pump flow returned the actuating cylinder's hydraulic fluid flow to the reservoir at pressures below 100 psi.

The booster system solely supplied power to the aileron booster unit, whereas the hydraulic pump, similar to the pump used in the main system received fluid from the air-pressurized reservoir. The main and boost system used as the reservoir was one package; however, the reservoirs operated as separate units and functioned independently of each other. The boost system pressure varied between 570 psi to 855 psi. Fluid from the pump flowed through the automatic relief valve, through the accumulator mounted on the left side of the fuselage at the wing root through a filter assembly to the boost disconnect valve, and to the booster unit. When the boost unit was not in operation, the fluid flow from the pump returned to the reservoir at a low pressure, below 100 psi.

The main hydraulic system performed no flight essential functions and losing the entire system failed to lose the airplane the exploitation team the plane used back up pneumatic sources for lowering the landing gear and flaps should the hydraulic system become in operative. The aileron control system reverted to manual control if the hydraulic boost system failed, providing safety of flight condition depending on the airplane speed and attitude. The rudder and elevator surface controls manually operated through push-pull control rod mechanisms. Removal of the hydraulic pumps from the engine appeared to be a very difficult and time-consuming task with the boost system pump mounted directly above the main system pump with little clearance available through the fuselage access holes to apply a wrench to the pump mounting bolts. In general, the hydraulic and pneumatic system components were heavier and bulkier than similar components used in US military airplanes.

The hydraulic system design was not readily vulnerable when subjected to conventional gunfire damage. The pilot could still fly the airplane to safety with the systems completely disabled; however, the hydraulic fluid used in this airplane was quite flammable and, if ignited, capable of creating extensive aircraft damage. Since most of the fluid in the system was located forward of the hot sections of the engine, leaking fluid could readily ignite by the cooling air carried it into the hot section of the engine and be ingested it into the engine compressor. However, this area also contained the engine fire extinguishing system to abate any hydraulic fire and associated fire damage.

The FRESCO C pneumatic system provided power for emergency lowering of gear and flaps, main gear wheel brakes, gun charge, cockpit canopy seals, electronic component pressurization and canopy deicing. The plane stored its air supply in two cylindrical air tanks, each with a volume of 244 cubic inches at a pressure of 2140 psi, and a spherical flap emergency tank, volume 122 cubic inches at a pressure range between 1560 to 1850 psi. These tanks were installed in the nose wheel well of the airplane. Air was also stored both in the landing gear strut for the purpose of emergency extension of the gear and emergency wheel brake operation and in the armament service tank for gun charging and wheel brakes. Air was stored in the strut and armament tanks at 710 psi with a volume of 402 and 122 cubic inches, respectively. Air from the main tanks supplied the pressure regulator and the gun charge, wheel brakes, canopy seal and deicing and the electronic pressurization systems. The gun charge used air at 710 psi, and the brakes from 57 to 100 psi. The remaining main tank systems used 40-psi air. The plane obtained its emergency air pressure for wheel brakes from the spherical tank. The pneumatic system charged externally at a single point connection.

Fracturing any one of several lines located near the right side of the cockpit could disable portions of the pneumatic system, a problem because gunfire could easily hit the 1/4 inch in diameter pneumatic aluminum alloy lines. The same applied to the main tanks and spherical tank pneumatic lines extending along the bottom portion of the right fuselage about two feet below the leading edge of the wing root.

From a vulnerability viewpoint, gunfire damage to the pneumatic lines rated more severe than damage to any of the three pneumatic bottles based on the assumption of the steel bottles not exploding and distributing shrapnel. The US military required steel air bottles to pass a gunfire test to demonstrate this non-explosion capability, and the exploitation team assumed the pneumatic bottles used in the MiG-17F could meet this requirement. The exploitation team also found the hydraulic and pneumatic lines extending down along the front of the landing gear vulnerable to foreign object damage when airplanes took off in close formation.

During this evaluation, the exploitation team removed the aft section of the aircraft fuselage to expose the installation of its hydraulic components; however, the exploitation team did not make or perform any tests on these systems other than that required to check out the system's operational suitability.

Evaluating the electrical power system included the major components of the electrical power generation and distribution system including the generator, voltage regulator, power relays, reverse-current relay, battery, and inverters. The exploitation team visually inspected all major components of the electrical system in sufficient detail to identify system functions and to determine the significant technical details of each component.

The electrical system was a simple single-generator DC system rated at 6 Kw (200 amp) with a battery backup nearly identical to US direct-current systems used for the past 20 years. The exploitation team found the generator, voltage regulator, and reverse-current relay inaccessible for maintenance without pulling the aft fuselage section.

An 18-cell, 28 ampere-hour nickel-cadmium type battery mounted in a thermally insulated case with the individual cells with metal cans encased in a thin, clear, plastic insulating material. A one-piece molded plastic grid spaced and supported the cells in the battery. The molded plastic cell filler and vent caps were as advanced as present-day US practice, indicating the battery design not being nearly as old as the aircraft, and probably replacing the original battery to obtain better low-temperature starting capability.

A separate electric starter motor provided for engine starting, incurring a weight penalty of approximately 30 pounds over the combination starter-generator unit used in current US aircraft.

The battery provided an emergency source of electrical power during any loss of the engine-driven generator system. The battery could supply all electrical power requirements for an estimated period of 15 to 20 minutes. Connecting only the essential loads doubled this length of time. Other than air starts, losing the battery did not cause loss of electrical power as long as the generator system remained functional. Loss of the complete electrical system caused loss of avionics, radar, communications, starting for engine and related systems, ignition, fuel boost, fire warning and extinguishing, and instruments except tachometer and temperature.

The exploitation team considered critical to the electrical system the inverters located in the cockpit at fuselage frame 9 and in the gun package well at frames 5 and 9, and the voltage regulator and reverse-current relay located on the right-hand side of the bulkhead at fuselage frame 13. The exploitation team also considered critical the battery located in the upper part of the fuselage nose at frame 2, the generator mounted on the upper left-hand part of the engine accessory section at fuselage frame 15. Naturally, this included switches, relays, fuses, and circuit breakers distributed throughout the cockpit.

The overall effect of the loss of the electrical system would render the aircraft completely ineffective as a weapon system, however, would probably not prevent the aircraft from returning safely to base.

Visually examining the fire and overheat detection, the exploitation team determined the type of detectors or sensors used, and the physical location of the system with respect to areas most vulnerable to temperature extremes where danger of fire existed. The MiG-17F utilized four thermo overheat detectors with temperature alarm settings to exceed the normal ambient temperatures of the protected area.

The MiG-17F used a conventional one-shot fire extinguishing using two 3-litre chromansyl cylinders. Each of carbon dioxide gas under pressure of 225 Kg/Cm mounted at frame 18 behind a fire shield with a pipe connection running along side the engine to the manifold where the discharge ring mounted on the engine around the jet turbine body at frame 19. The discharge valves on the cylinders were cartridge actuated and fired simultaneously by depressing a fire extinguisher button on the left control panel in the cockpit. The extinguishant discharge directed forward into the engine compartment. The circuitry for the fire detecting and extinguishing system took from two silver zinc batteries located in the nose compartment forward of the cockpit.

This aircraft cooling system design took advantage of ram air entering through the split engine inlet duct for functioning all the necessary cooling. A duct vent cooled the alternator and another cooled the engine accessories by forcing cooling air behind a shroud around the turbine section. An engine-mounted shroud utilized around the afterburner cooled the actuator nozzles and tail pipe.

The exploitation team found that a hit could damage the hydraulic lines and pump the fluid into the hot area where a fire was bound to occur. The aircraft had an emergency fuel shutoff system but no means of shutting off the hydraulic system.

A hit occurring between frames 17 and 19 damaging one or more of the fire detector might not alarm the pilot to initiate emergency procedures. It appeared that the extinguishing system designed combated only ground fuel fires.

The deicing system utilized on the canopy windshield did no afford ice protection to the leading edges of the aircraft. The system operated off the main pneumatic (air) system and used Ethyl alcohol for antifreeze fluid eject by the pilot using an electric push button on the left control panel in the cockpit. The circuitry for the ice removal system came from the batteries located in the nose compartment forward of the cockpit and enabled the pilot to switch the alcohol supply on from 10 to 18 times, each time for 2 to 3 seconds, and to remove ice to 3 mm thick.

The MiG-17F airframe used a simple fuel system consisting of two fuselage and two external tanks. The Number 1 fuselage tank located between the cockpit and the engine carried approximately 330 gallons. The tank, bounded on the sides by the air ducts, used a non-collapsible, bladder type liner to fill the cavity. Fuel pumped from the compartmented sump tank directly to the engine. The Number 2 fuselage tank, all metal tank and located around the engine, held approximately 43 gallons that pumped to the Number 1 fuselage tank. The aircraft carried two jettisonable external fuel tanks with a capacity of approximately 105 gallons each. The external tank fuel transferred to the Number 1 fuselage tank by air pressurization. The exploitation team found nothing unique or outstanding about the fuel system except for its simplicity due to the limited fuel capacity of approximately 500 gallons and having good vulnerability design protected by the engines. The cockpit area bordered the tank on the forward side. The exploitation team found no critical components in the fuel system except possibly the Number 1 tank boost pump, which the exploitation team believed boost failure failed to keep the aircraft from operating at some reduced altitude or power setting.

In evaluating the life support systems, the exploitation team examined the MiG-17F oxygen system, restraint harness, cockpit layout, visibility, egress equipment, descent and survival equipment, and design philosophy. The exploitation team found the personal equipment and cockpit layout comparable-to western equipment developed during the same period.

The exploitation team found the cabin small and narrow with the ejection seat very snugly fitted into it consistent with an aircraft with a low frontal profile design. The distance from the outside of one, 1-inch thick, ejection handle to the outside of the other (seat width) was 19 inches and normal for a fighter ejection seat where narrowness was paramount. The cabin cramped pilots taller than six feet and heavier than 200 pounds, however, ninety percentile and below of Caucasian pilots fitted into the seat with ability to control rudder pedals effectively.

The presence of the gunsight, reticle, and associated equipment restricted forward visibility, however lateral and posterior visibility appeared good. A rear view flat mirror periscope with a miniature blind did an excellent job of preventing bright sunlight reflection. The pilot increased visibility slightly by placing padding under the buttocks, raising the sitting height limited by helmet contact with the canopy.

Similar to early vintage western interceptor and fighter jet aircraft, the MiG-17F lacked protection for flash blindness.

Surprisingly, the MiG-17F did not carry a survival kit. It provided for the storage of survival equipment between the parachute and the buttocks pad in an empty space measuring about 14 x 15 x 5 inches, a useful place to put a life raft (deployable after removal of the restraint harness) and/or survival gear, radios, flares and the like.

The ejection seat copied a dual motion, early vintage, ballistically lofted Martin-Baker seat. It lacked the later additions that USA and NATO applied to their seats to improve ejection success rate. It also lacked a zero delay lanyard, and its equivalent brass key (time to seat) appeared unreliable and easy to accidentally disconnect. Ejection required three motions rather than the desirable single one, making mid-air automatic or manual deployment of the survival-flotation equipment difficult or impossible due to central metal reinforcement. Padded leg retainers prevented leg flailing at high speed by locking the pilot's heels in stirrups that released after ejection by a timer. The torso harness and quick release box deserved d considerable mention due to their tried and true design and the active US DOD/industry interest in this area. The torso harness design sewn to the parachute usually remained in the aircraft for quick reaction alerts and convenience for pilot who could very easily don the parachute unassisted (10-20 seconds for the trained crews). The four ends of the torso harness ended in a single simple two-motion release box often called a quick release box located in an easy place for release, by either hand, in the front lower chest area.

The chute D-ring was in cloth housing with elastic retainer. The exploitation team pull-tested two different D-rings at twenty pounds US mil spec. Rather than liquid oxygen, the oxygen system used three high-pressure-2250 psi gas cylinders and a Diluter demand regulator using a knurled knob. When manually opened, it delivered pure oxygen at manually selected pressures up to 44 mm Hg, which was satisfactory for a few minutes at 45,000 feet without cabin pressure. A seat pan in the chute contained about one liter of oxygen at 2100 psi that automatically activated upon ejection.

Evaluating the flight control system, the exploitation team found the primary flight controls having a rigid push-pull design without any duplicate rigid or cable systems. The primary flight control systems consisted of an Elevator Control System, Rudder Control System, and an Aileron Control System. The Secondary flight control systems were the Elevator and Aileron Trim Tab Control System, Flap Control System, Speed Brake Control System, and Landing Gear Control System.

A control stick in the cockpit and a series of control rods and bellcranks under the cockpit floor and along the top right hand side of the fuselage operated the elevator control. The control rods turned upward through the lower vertical fin where they connected to two rods, one going to each elevator between the two vertical sections.

Rudder control operated by the foot pedals in the cockpit and by a series of control rods and bellcranks under the cockpit floor and, then, along the top left hand side of the fuselage. The control rods connected to the rudder control tube in the lower vertical fin. A tube with a universal joint connected the upper and lower rudder sections.

The aileron control operated by the control stick in the cockpit and controls rods to a hydraulic boost in the right hand wing root aft of the front spar. A series of control rods and bell cranks through the wings controlled the ailerons. The aileron control was the only boosted flight control in the aircraft, however, if the boost was out or the lever in the cockpit was turned off the aileron could still be operated by applying an increased effort to the control column.

Switches in the cockpit electrically operated the elevator and aileron trim tab controls. The trim tabs were located on the left control surfaces only.

The flaps operated by a lever on the pilot's left hand panel, which connected to a hydraulic valve. This valve controlled the hydraulic actuators in the wings, which lowered and retracted the flaps. Two control rods operated each flap interconnected by a cable system, which synchronized the lowering of the flaps.

Hydraulic actuators controlled by a switch mounted on the pilot's left hand panel operated the speed brakes, while hydraulic actuators controlled by a lever on the left side of the instrument panel operated the landing gears.

Compressed air activated the emergency landing gear operation, extending all gears or only the nose or main gears. For emergency extension, the pilot manually released the uplocks by pulling the handles for each main and nose gear. The main gear had their own emergency compressed air supply located in the main gear legs that the pilot used to lower the flaps to the full 60 down position in case of a hydraulic failure.

An interesting part of the exploitation of the Soviet planes focused on materials, manufacturing technology, corrosion protection, and manufacturing methods. The exploitation team examined the airframe structure, landing gear, and engine with the aircraft broken at frame 13. The exploitation team found the Soviets used WWII techniques in their manufacturing processes and fabrication procedures for the MiG-17, much as they did with the HAVE DOUGHNUT MiG-21. The exploitation team observed no unusual manufacturing or fabrication procedures and good workmanship in general with the basic material being-2024 type aluminum alloy, and Chromasil steel for the manufacture of the landing gear carry-through structure and the primary load-carrying joints.

Steel armor plate approximately 0.40 in. thick forward of the instrument panel, and 0.641 in. thick armor aft of the pilot's seat up to the pilot's head protected the pilot from the front and rear. The canopy contained a steel armor headrest and bulletproof glass using laminated silicate for the pilot's protection from rearward facing armaments used in US bombers of the middle and late 1950-1960 period. The exploitation team found the primary construction material in the MiG-17F unlikely to fail catastrophically from a single direct hit other than large caliber hit directly on the wing root or load carrying wing spar.

The exploitation team found the outside of the plane painted with no primer sprayed on the interior surfaces after final assembly. Corrosion protections appeared standard with some evidence of corrosion on the steel rivets and steel bolts.

The aircraft structure of the all-metal, mid-wing FRESCO C aircraft contained sweptback wings, empennage surfaces, and a tricycle design landing gear with brakes on the main gears. Its fuselage tanks and two under wing tanks were mounted just outboard of the main landing gear attachments with a maintenance break at bulkhead number 13 for engine access, dividing the fuselage into nose and aft sections. A compartment in the nose section and a pressurized cockpit housed radio and electronics equipment. The nose gear retracted forward in a compartment below the radio equipment. Aft of the nose gear, two 23 mm and one 37 mm cannons attached to the bottom of the fuselage. The plane's speedbrakes were located at the end of both sides of the aft fuselage section.

The wing swept back at an angle of 45° along the aerodynamic center. The nose section contained the internal aerodynamic balance weights and power boosted controls for the ailerons. The flaps moved to two deflected positions, 20° for takeoff and 60° for landing. The empennage consisted of a fixed vertical stabilizer with upper and lower sections, and an upper and lower rudder section. The horizontal stabilizer and elevators were located about midpoint on the vertical between the up per and lower sections.

The pressurized cockpit came equipped with an armor plate forward of the instrument panel and an ejection seat with a small piece of armor plate attached to the seat back. The fuselage skin was .054 inches thick aluminum, designed in two parts, a forward and aft section, which permitted quick disassembly for engine maintenance.

The nose section included 13 bulkheads from the nose cone to the fuselage break, three partial bulkheads, four longerons, and stringers. The nose gear attached to the nose section at bulkhead number 4 and retracted for ward into the lower fuselage to bulkhead number 1.

The radio and electronics compartment was directly above the nose gear location. The pressurized cockpit was located between bulkheads number 4 and 9 in the upper nose section. The cannon support structure attached to a steel fitting just forward of bulkhead number referred to as 5A, and to a steel fitting on bulkhead number 9 on the lower fuselage. Bulkheads numbers 1, 4, 5A, 9 and 13 were the main load attaching points in the fuselage nose section. An external stainless steel doubler riveted to the nose cone and aft to bulkhead number 3 for cannon gun blast protection. The skins and fasteners on the fuselage nose section were of good aerodynamically smooth workmanship.

The cockpit floor constructed of sheet aluminum with riveted connections was located between the inner duct wall structure and extended from bulkheads numbers 4 to 9. A fuel tank was located between bulkheads numbers 9 and 13 and was in stalled and removed through a door on the lower fuselage which attached by steel bolts. The aft fuselage skin was aluminum and had access doors for engine equipment and flight control systems.

Evaluating the empennage, the exploitation team found the vertical tail separating the two vertical sections constructed in lower and upper fin and rudder sections. The upper vertical fin was removable for removal and replacement of the horizontal tail. The lower vertical fin permanently attached to the aft fuselage section. The lower vertical fin construction included one main spar, a front and tailing edge spar, an intermediate spar, stringers, and ribs. The rudder consisted of a main spar, nose spar ribs and a trailing edge extruded or machined member. The frame and skin were aluminum connected by aluminum rivets with the mass balance weight of the upper rudder extending into a cut out in the tip of the vertical fin. The rudder operated by a rigid control rod system connected to the lower rudder within the fuselage tail assembly. A rod with a universal joint connected the upper and lower part of the rudder.

The right and left hand parts of the horizontal stabilizer rigidly attached at the centerline in a main spar, rear spar, stringers, and rib construction design. The main spar was steel at the inboard end however; it could not be determined how far outboard the steel extended. The main attachment was to the lower vertical fin main spar by steel fittings and bolts. The horizontal did not have a front or nose spar, as did the other lift surfaces.

The right and left hand designed parts of the elevators were not interconnected. To control each elevator, a separate control rod connected to a single control rod system in the lower vertical fin. The elevators suspended on three hinge points with the center hinge being a floating type to assure proper alignment. A mass balance weight was located in each elevator near the tips and extended into cutouts in the horizontal stabilizer. Only the left elevator possessed a controllable trim tab.

The FRESCO C was equipped with a tricycle type landing gear. The nose gear had a level wheel suspension, shock strut with a lower offset fitting which provided the pivoted point for the lever type wheel fork. The main gears were also a lever wheel suspension, but with no shock strut incorporated in the main gear member and no transfer of the bending to the fuselage structure. Only the main gear wheels had brakes installed. The FRESCO used tube type tires of two sizes, main gear-660 x 160 mm, and the nose gear-480 x 200 mm.

The wing-mounted main landing gears pivoted on a steel fitting attached to the front spar and a forged boss on the main beam. The gear, held by an uplock assembly, retracted inboard and aft into the wing. Three doors, the center door attached rigid to the gear and the other two hinged to the wing, covered the opening in each wing. A hydraulic actuator operated the inboard door and the outboard door by a link attached to the gear. The main gear was held in the down position by the retraction strut which also was the drag and side load member. The main gears were constructed of small steel forgings welded together to formed the complete gear. Attachments for such as the retracted strut, doors and servicing port welded to the gear assembly. The axle had both inboard and outboard brake drum backing plates, which provide a double brake for each main wheel. The plane used the landing gear leg inner cavity as a tank for compressed air for emergency lowering the main landing gear. The exploitation team never identified the material of the wheels; however, believed it to be magnesium.

The nose gear mounted in the forward fuselage structure and retracted forward into a well in the fuselage as an assembly constructed of small steel forgings welded together. The wheel fork was a Y design and was a weldment of small parts with a small vibration or shimmy damper attached on the aft side of the shock strut member. A self-contained compressed air emergency system for lowering the nose gear was independent of the main gear system.

In their vulnerability analysis of the MiG-17, the exploitation team denoted the characteristics of the MiG-17F system, which caused it to suffer a finite degradation (in capability to perform the designated flight) because of out subjecting it to a certain level of effects in an unnatural hostile environment (manmade). The exploitation team defined threats as ranging from a single gun-fired projectile to fragmentation and internal blast from a high explosive warhead. The exploitation team did not consider external blast being a lethal threat because the lethal radius from the weapons considered was less than that of the fragmentation. The exploitation team defined critical components as those system elements necessary to maintain controlled flight.

There was little or no masking protection except for the wing or inlet ducts and no armor on either side. There was no masking or armor for the top and bottom unless one considers the gun system as a masking component. The gun system was not necessary to flight but could be extremely hazardous from fire and explosion.

Non-critical elements and with an armor projected to defeat a .50 caliber threat masked the front of the cylinder. The 50-caliber gun as tail turret armament for the FRESCO-C was typical of that used by US bombers and air superiority fighter armament at the time.

The engine tail pipe and aft fuselage masked the back of the above cylinder. The FRESCO C power plant was of radial or centrifugal compressor design with dual could combustion chambers, a single turbine and after burner. There were several vulnerability studies on this engine and the general consensus of opinion was that the above type engine was not susceptible to one hit kills from 50 caliber type munitions. The radial flow (centrifugal) compressor was much less vulnerable than the axial flow compressor. The taper in the tail pipe, tail pipe nozzle, afterburner, and aft fuselage masks the turbine blades and diffuser nozzle from the rear. The power plant and above elements masked the critical component from the rear. The empennage and aft fuselage were not extremely vulnerable from any direction except for about one square foot of area covering the control element at the root of the horizontal stabilizer in the vertical stabilizer.

The pilot's head and shoulders extended above the masked protected areas. Flat plate armor protected the pilot's shoulders from the rear, while armor curving around the headrest protected the back of the pilot's head; however, heavy glass in the front did little to protect the pilot from a 50-caliber threat.

The FRESCO C weapons system could well absorb single hits from existing domestic gunfire threats from the rear unless they hit a control rod or deflected into the turbine blades and diffuser nozzle. However, a short duration burst from the presented domestic air superiority gun threat at ranges to 3000 feet could well cut the FRESCO in half, which would undoubtedly constitute a kill. In general, the presented air superiority missiles would kill the FRESCO if they detonate on contact or near proximity (10 feet) with the aircraft. The front and rear of the vulnerable cylinder extending from the engine accessory section to the front of the crew station in the FRESCO C were masked as well as possible. The exploitation team concluded that the area presented for possible one-hit kills decreased as a function of the cosine of the angle as the angle of attack varied from normal to the fuselage axis. The philosophy of designing the FRESCO C was to mask the critically vulnerable components of the system from the prime directions of attack (front and rear) with armor or non-critical and/or hard components to defeat the existing threat for the design period.

In evaluating vulnerability of elements critical to flight (airframe), the exploitation team found the airframe very hard with only three points constituting one-hit gunfire kills. These points were the wing roots, which were massive and very difficult to destroy. The exploitation team also found the empennage control section susceptible to a one-hit kill. The structure around the main fuel tank and accessory section was very susceptible to secondary effects, such as fuel fire or hydraulic fluid fires and hydraulic ram damage. The FRESCO C total structure was vulnerable to the current missile threat given a hit or close proximity detonation.

Evaluating the pilot and crew station, the exploitation team found the pilot well protected from the front and rear by masking and armor as described in the exploitation. The gun platform and magazine protected the pilot from the bottom, but he was virtually unprotected from the top and sides. The pilot was a critical element. The windshield was supposed to be some type of bullet proof glass and it being impossible to tell exactly what threat level was would defeat unless it was subjected to a series of threats.

The exploitation team found the FRESCO C hydraulic and fuel pumps, generator, and air compressor masked front and rear, but unprotected from the sides, top, or bottom, making them vulnerable to a 50-caliber threat.

The taper in the tail pipe, the aft fuselage, and afterburner nozzle masked the engine turbine blades and diffuser nozzles. The solid portion of the engine turbine wheel was a hard target that masked the engine from the rear and from the front. The inner liners of the engine combustion chambers made them insusceptible to one-hit kills from anything less than an explosive warhead. The exploitation team also found the radial type (centrifugal) engine compressor insusceptible to self-destruction by progressive damage from one-hit by a 50-caliber threat.

The engine had many critical areas where a 50 caliber API round or a high velocity (4000 fps).100-200 grain fragment could destroy the system balance or start a fire. However, the vulnerable components on the FRESCO C engine proper presented a very small target area. The loss of the engine in the FRESCO C did constitute a kill since it was a single engine aircraft.

The loss of the electric generator limits the operation of the aircraft to battery power. The loss of battery power and not the generator would destroy air restart capability. The loss of the hydraulic system would not kill the aircraft flight capability because the FRESCO C did not require hydraulic power for flight control actuation. However, fires resulting from the hydraulic fluid could well destroy the integrity of the structure of other elements critical to flight. The hydraulic accumulators were positioned in the wing roots and constituted a threat from the standpoint of fire and explosion if hit from top and bottom.

The rear fuel saddle tanks located in the aft fuselage emptied shortly after take-off and did not constitute a fire threat. Their non-rigid construction eliminated them as a serious explosion threat. The wings on the FRESCO C were dry.

The main fuel tank was located between the pilot and engine accessory section. It was well masked from the front and rear, but did constitute a serious threat if hit by the proper threat. Hydraulic ram and fire from the main tank would result in catastrophic failure of the aircraft in general in the area of the tank.

The critical flight controls were primarily mechanical and thus presented the smallest possible vulnerable area. The only power actuators in the flight control system were the aileron boost, and something that the exploitation team did not consider critical to flight.

The-munitions magazine could well create a fire or explosion hazard to the aircraft, but it was not a proven fact. The above element could well provide protection to the pilot from the underneath side.

Masking of well-designed compressed air bottles positioned critical components prevented their creating an explosive hazard.

Fire extinguishing bottle-the fire extinguishing bottle, if hit, could saturate the engine inlet with CO_2 for a short period and starve the combustion process of oxygen or destroy critical accessories if it explodes,

The exploitation team did not have pressure bottles available for gunfire tests to establish their vulnerability to fragments and conventional projectiles.

The philosophy of aircraft survivability design for the FRESCO did not use active survivability counter-measures other than the radar warning in the tail and non-smoking fuel combustion. The FRESCO C design concentrated the vulnerable critical flight components in a small volume thus presenting a small vulnerable target. Also, the survivability design philosophy protects these vulnerable components by masking and armor from front and rear. There was no evidence of fire and/or explosion retardants in the-fuel system. The plane used the engine accessory section fire extinguishing system primarily for ground fires since it was a one shot affair and not designed for use on an engine fire in flight.

The exploitation team concluded from their evaluation that the FRESCO C was a hard system for the threat during the period of its design. The FRESCO C could well absorb single hits from the current gunfire threats but not multiple hits.

The FRESCO C was a well-designed, simple system for its period of design. Its simplicity allowed full maintenance by underdeveloped nations.

The exploitation team evaluated the scan-fix radar to determine operational characteristics and radar susceptibility to jamming. The equipment appeared to be of good construction and design corresponding to the early 1950s.

The exploitation team found that the exploitation team could change the receiver-transmitter unit in approximately 10 minutes, and thus, could change it during aircraft turn around time. The PRF (Pulse Repetition Frequency) was adjustable by a potentiometer in the receiver-transmitter unit, but it required removal of the unit and a bench adjustment.

The transmitted radio frequency was not adjustable, and any change of magnetrons to accomplish a frequency change would necessitate complete receiver tuning of the Klystron, transmit-receive (TR) tube, and the crystal mixer assembly.

There appeared to be no ECCM fixes incorporated. However, perhaps one could consider the AGO noise threshold, the video integration, and the gate-position memory a formed of ECCM.

The exploitation team verified that the target abandonment junction required 3 to 4 seconds to discard a target under track. The exploitation team expected this memory time adjustable via a potentiometer marked "memory time setting" located in the range unit.

The exploitation team observed that after target abandonment, the range gate appeared to drift from maximum range towards minimum range, and would lock up on a target while drifting toward minimum range. Therefore, the magnetron could cause a variation in radar range due to the consequent variation in starting pulse amplitude or width.

The procedure to adjust the various potentiometers in order to obtain suitable radar performance was lengthy because of the interaction of adjustments, some of which were quite sensitive. In several cases, the exploitation team interchanged or replaced tubes by tubes extracted from the spare unit in order to improve the radar performance.

In conclusion, with regard to the difficulties encountered, the exploitation team consumed approximately 75 percent of their time just correcting malfunctions necessary to maintain the radar at a level necessary to provide expected radar performance during the time the exploitation team had the radar available for exploitation.

In view of the numerous difficulties encountered to maintain the radar in operation, it was quite conceivable that some of the data collected not might be representative of the SCAN FIX radar encountered in the field. The exploitation team reached the following conclusions based solely on the one radar examined:

- Radar performance indicates that sensitivity-time-control (STC) was employed. The circuit schematics do not obviously support this conclusion.
- Range gate stealers that reduce the apparent range require less power to capture the range gate than range gate stealers that attempt to increase the apparent range. Generally, range gate stealers attempt to increase the apparent range.
- A jamming to signal (J/S) ratio of 10 db was sufficient to break-lock and to deny reacquiring the target in the range gate.
- The deployment of chaff that blossoms no more than 50 meters from the aircraft should capture the range gate and break-lock on the dispensing aircraft.
- The radar provision for target abandonment allowed the range gate to drift from maximum range. This feature could allow the radar to consistently lock on a target, which was farthest from the radar, for example, a chaff dispensing aircraft.
- The minimum required power to lock on a target is-61 dbm. Break lock occurred at a 63-dbm signal level. The range gate memorized and maintained the last intercept range position for 3 to 4 seconds in cases where target signal fading occurred. The exploitation team acquired or processed no velocity information.

The following is a list of the major test equipments utilized in the tests:
- 1 TS-1011UPM-84 spectrum analyzer
- 1 EH-121 pulse generator
- 3 EH-131 pulse generator
- 1 RF Generator
- 1 Polarad S-band RF generator-HP 616A
- 1 Telonic signal generator (nautical miles 2000 with an SH-1 plug-in head)
- 1 Tecktronic 555 oscilloscope
- 1 Teckronic 535 oscilloscope
- 1 Hewlett Packard power meter
- 1 Simpson Model 260 multimeter
- 1 Hewlett Packard vacuum tube voltmeter

CHAPTER 5- Performance, Stability, and Systems Evaluation

HAVE DRILL/HAVE FERRY AIRCRAFT

The AFFTC and NATC customers, in conjunction with the Special Projects exploitation team, conducted the performance, stability, and systems evaluation of the MiG-17F at Groom Lake.

Following the technical exploitation, the HAVE DRILL MiG-17F made its first US flight at Groom Lake in January 1969. The second MiG-17F codenamed HAVE FERRY flew in March of that year. Flying both planes, the exploitation team evaluated the performance, handling qualities and systems evaluation of the MiG-17F exploitation programs led by AFFTC and NATC. The exploitation team (the exploitation team as a whole) flew the two airplanes for test a total of 20.7 hours in 33 flights, equipping only the #1 airplane with test instrumentation for obtaining quantitative data. Therefore, all data presented herein refers to the #1 (HAVE DRILL) MiG-17F unless otherwise stated.

The exploitation team, consisting of CIA, Air Force, Navy, and Special Projects contractor personnel, obtained the quantitative data using a clean airplane (no external stores) and limited quantitative data with two external fuel tanks (106 gal/tank). The engine start gross weight with no external stores was approximately 11,700 pounds and with two external fuel tanks was approximately 13,200 pounds. No attempt was made to vary the center of gravity so handling qualities at extreme g were not tested.

The exploitation team found the MiG-17F a simple, sturdy, inexpensive, easily maintained aircraft with excellent maneuverability capabilities at 450 KIAS and below. Canopy visibility was excellent except out the front horizontally and also forward and down. The periscope mounted on the canopy was a simple, but excellent aid to the pilot. It provided rear hemisphere clearance for the pilot with minimum head movement. This was particularly important in the high g maneuvering engagement. The simplicity of the aircraft and its systems made it possible for US pilots to start flying the aircraft at maximum performance after only two checkout flights.

The MiG-17F airplane was a single-seat, swept-wing, point defense interceptor powered by a VK-IF centrifugal flow jet engine with afterburner. The VK-IF engine had an uninstalled sea level static thrust rating of 7,440 pounds at maximum afterburning and 5,720 pounds at military according to the manufacturer's data. Without stability augmentation, the plane powered only the ailerons of the primary control surfaces. Its primary control linkages were rigid push/pull tubes in the longitudinal, lateral, and directional axes. It carried its fuel in two internal fuselage tanks of 1250 and 16 0 liter capacity and, optionally, from two external wing tanks of 400 liters each.

Measured parallel to the chord, the symmetrical wing had an average thickness ratio of 8.8 percent and a 45-degree sweepback of the quarter chord. The leading inboard edge sweptback 49 degrees with an outboard sweepback of 45 degrees. The wing had a dihedral angle of minus three degrees and an incidence angle of one degree at the root. On each side, two full chord and one three-fourths chord stall fences attached to the upper wing surface. The ailerons occupied the outer one-third of the trailing edge of the wing with internal aerodynamic balancing with a fabric sealing the gaps between the wing and ailerons. Although the pilot controlled the ailerons manually with boost off, the lateral control system incorporated a hydraulic booster with a 23.4 boost ratio and specified maximum aileron travel of 18 degrees trailing edge up and trailing edge down (TEU and TED). An electrically operated lateral trim tab was located on the left aileron and had a maximum travel of 15 degrees TEU and TED.

Split flaps with a sliding axis of rotation covered 47 percent of the inboard trailing edge of the wing. When operated, the flaps moved slightly backward and simultaneously downward. The pilot normally operated the flaps hydraulically but could operate them by a pressurized air system in an emergency. Flaps remained up, deflected, 20 degrees for takeoff and always 60 degrees for landing.

The horizontal stabilizer mounted 60 percent high on the vertical fin had a 45-degree sweepback with an incidence of 0 degrees 33 minutes, which could be set only on the ground. A manual reversible longitudinal control system operated the mass-balanced elevator mounted on the horizontal tail. Maximum deflections of the elevator were 32 degrees TEU and 16 degrees TED. An electrically operated trim tab located on the left elevator had a maximum deflection of +10 degrees.

The vertical stabilizer had a 55 degree 41 minute sweepback. A manually operated mass-balanced rudder consisting of two sections, one above the horizontal tail, and one below deflected a maximum of 25 degrees trailing edge right and trailing edge left (TER. and TEL) with trimming by means of a bend tab changeable only on the ground. The two hydraulically operated dive (speed) brakes mounted on the sides of the aft fuselage deflected a maximum of 55 degrees.

The VK-lF engine used a single stage double-faced centrifugal compressor with combustors consisting of nine cans. The turbine was a single stage, axial type, and the exhaust nozzle a hydraulically operated two-position convergent nozzle with a maximum afterburning diameter of 24.38 inches and nonafterburning diameter of 21.75 inches. Maximum afterburner and military engine speed was 11,560-20 rpm with an exhaust gas temperature of 718 degrees C. Engine speed for minimum afterburner was approximately 10,870 rpm. Changing engine rpm accomplished thrust modulation between minimum and maximum afterburner.

Each airplane carried one 37 mm gun, two 23 mm guns, and a gun camera. The 37 mm gun fired 40 rounds at 400 rounds/minute, and each 23 mm gunfire 80 rounds at 900 rounds/minute.

For the evaluations, the exploitation team instrumented the HAVE DRILL MiG-17F for performance and handling qualities tests, and instrumented the HAVE FERRY MiG-17F only with several sensitive cockpit instruments.

The cockpit of the aircraft was an antiquated design impossible to enter or exit the cockpit with any degree of urgency. It required the pilot to step on the seat type parachute, which was an integral part of the seat, supporting himself with his hands on the canopy rails while threading his feet onto the rudder pedal stirrups. It required an average of two minutes to don the parachute harness and hook up the necessary personnel leads after entering the cockpit.

Pilot restraint mechanism consisted of a semi-conventional lap belt and shoulder harness arrangement. The pilot found it difficult to clinch the lap belt tight due to an awkward out and downward movement of the tightening straps. The shoulder harness pulled tight in a conventional manner but failed to stay tightly adjusted under normal pilot movement. The shoulder harness locked or released by movement of a locking knob located on the forward left side of the seat. The locking knob was opposite to USAF standards, moving forward-to release the shoulder harness, while movement aft locked it.

The personnel leads (oxygen/communications, etc.) were an integral part of the seat. Seat height was not adjustable and required padding for varied sitting positions and percentile persons. The rudder pedals adjusted manually prior to entering the cockpit but the pilot could adjust them after entering the cockpit. Pilots over 6 feet tall and husky in stature found the cockpit, the approximate size of an F-86F with less leg room, small, cramped, and difficult comfortably to fly the aircraft.

Foreign Technology Division cockpit evaluation

FTD evaluation of areas which limit or enhance cockpit visibility as viewed from the pilot's seat of the MiG-17F occurred through ground measurements of the cockpit visibility were made in a hanger using fixed reference points to measure and then calculate all angles.

A pilot sat in the cockpit simulating flight conditions by wearing a helmet, oxygen mask, and parachute with the pilot's safety belt and shoulder harness securely fastened.

The MiG-17F canopy provided good visibility above the horizontal plane. The gunsight restricted forward flight visibility through the windscreen and required pilot head movement to circumvent this obstruction. The canopy framework, antenna strips, and mounting assembly of the aft looking periscope slightly restricted upward, and the ejection seat headrest restricted visibility rearward.

Side visibility below the horizontal plane was restricted to approximately 40 degrees. This was due to the low pilot seating position and bulbous nature of the fuselage. The large size and position of the wings further degraded visibility to the aft and low positions.

The canopy-mounted periscope was an excellent aid to the pilot providing rear hemisphere clearance with minimum head movement. The canopy limited side-to-side head movement and the headrest limited aft head movement.

GUNSIGHT RESTRICTING WINDSCREEN VISIBILITY (S/NF)

WINDSCREEN - OUTSIDE VIEW (SECRET)

MiG-17F profile

This section reports the results of the exploitation of the HAVE DRILL vehicle, with respect to its weight, balance, and inboard profile data. This exploitation did not disclose any major deviations from previous information on the aircraft weight, center-of-gravity and inboard profile. The major contribution of this effort was the expansion of FTD's target description database with respect to subsystem and electronic component sizing, weight and positioning data.

AFFTC weighted the aircraft in a level flight attitude with no crew, full fuel and oil, and test instrumentation of photo panel, oscillograph package, inverter, and photo panel camera with gear down.

They weighed the plane using the Cox & Stevens Electronic Weighing Kit Type C-l, and Howell platform type weighing scale by placing the scale under the nose wheel and a jack (with electronic cell) at each wing jacking station. They weighted the airplane following the assembly and instrumentation to determine the center of gravity and ballast requirements, finding the specific weight of the fuel 6.39 pounds per gallon at the time of weighing. The fuel volume of 373 gallons resulted in a total fuel weight of 2,385 pounds.

The bubble canopy, a forward and aft sliding clamshell, manually operated with positive mechanical finger locks at the open and closed positions. To insure cockpit pressurization, the pilot actuated an inflatable rubber seal with the initial rotation of a pressurization/temperature mixer handle located on the right side of the cockpit. A rearward facing periscope mounted on the top forward edge of the canopy afforded a much superior rearward vision to the conventional rearview mirror used in US fighter aircraft. With the pilot's head in a normal erect position, he saw the entire horizontal tail through the periscope. Moving his head increased his field of vision to +40 degrees above and below the horizontal tail and 80 degrees side-to-side. A switch in the cockpit electrically heated the periscope for defog.

Cockpit field of view was fair; however, visibility was considerably restricted by the gunsight, forward canopy bow, periscope housing, and to some lesser extent by the integral radio antenna wires and distortion within the canopy.

The aircraft accommodated the installation of two different ejection seats, one version having armrest "triggers" that the pilot squeezed for both canopy, and seat ejection; the other, a "face curtain," and arm rest "triggers" for ejection. The face curtain was recommended as primary on airplanes where installed.

The pilot could readily reach all switches with the shoulder harness locked. During exploitation of the MiG-17, the exploitation team found the cockpit switching arrangement simple but adequate with the main electrical switches mounted in convenient groupings on both the left and right forward consoles. The exploitation team found the instrumentation generally of older design, but adequate for the airplane's flight. Most instruments appeared to be reliable and stable in their operation with the exception of the artificial horizon, which worked opposite to US designs in pitch and precessed excessively during hard maneuvering. Warning lights scattered indiscriminately around the cockpit lacked having any normal grouping. The production instrumentation lacked having a Mach meter, so the exploitation team installed one for flight tests.

Fuel quantity indication through a simple float/ potentiometer system mounted within the main fuel cell proved to be reliable but often resulted in erratic quantity readings due to fuel sloshing or changes in attitude. A red low level (300 liters remaining) provided a warning light in addition to the quantity gauge metric reading.

With minimal start procedures or cockpit checks, the engine started with no problem with either external electrical power or the aircraft battery. From idle rpm (2500) to approximately 7000 rpm, the throttle required carefully movement to avoid exceeding the engine's maximum temperature limits. The exploitation team experienced no problems above this range to the maximum rpm. A conventional throttle grip rotated for manual gunsight ranging contained the microphone transmitter button and the guarded engine crank switch controlled engine speed.

A four-position lever located just below the throttle quadrant hydraulically operated the flaps to lever positions that included UP, NEUTRAL, 20 degrees, and 60 degrees. The flaps reacted fast, requiring 1-2 seconds from full down (60 degrees) to fully retracted. This rapidity of operation coupled with the poorly designed flap lever caused loss-of-lift problems during critical phases of flight. The mechanized system prevented the flaps moving from full down to 20 degrees down without first moving the lever to UP and then to the 20 degrees down position. This required the pilot to fully retract the flaps and place them in the 20 degrees down position. In a heavy gross weight landing condition, the pilot initiating a go-around at low speed caused a large trim change, loss of lift, and possible loss of control by the momentary raising of flaps from full down to full up. The instrument panel indicated the selection of 60 degrees flaps by the protrusion of a small striped rod from the upper surface of the left wing and by a green light.

A small three-position lever on the left side of the instrument panel conventionally operated the landing gear, but included movement of the actuating handle back to the center NEUTRAL position once the landing gear traveled to the desired position. A guarded system POWER switch located just to the left of the landing gear handle had to be ON for any landing gear operation. This switch remained OFF for ground operation, which effectively prevented inadvertent landing gear retraction. Three green lights on the instrument panel and the extension of a striped rod on each wing and on the nose just forward of the canopy indicated extended and locked landing gears. Three red lights on the instrument panel and retraction of the indicator rods indicated landing gear UP.

The gunsight had both range only radar and manual ranging capability. In manual ranging the sight worked well while tracking a target up to approximately 3 g. When tracking above 3 g, the pipper jumped and jittered, making it difficult to track a target with any degree of accuracy.

The aircraft having no nose wheel steering made it necessary to used differential braking of the main wheels for taxi. Pneumatic air pressure stored in compressed air bottles and metered to the brakes by a hand-operated lever located on the control stick supplied brake pressure.

The pilot directed differential air pressure to the brakes through movement of the rudder pedals. The exploitation team found it best to taxi the MiG at a relatively high speed. Turns required full rudder deflection with metered air, so the exploitation team used small directional corrections with either full or small rudder deflections and metered air, depending upon pilot preference. The system essentially prevented small radius turns. Turns of 90 degrees, such as turning off the runway, required moderate speeds. Turns in excess of 90 degrees proved difficult, almost impossible for 180-degree turns within the width of a 100-foot wide runway. Directional pilot induced oscillation (PIO) could occur due to over-control while taxiing or braking. The most significant reason leading to this phenomena was the slight time lag required for the air pressure to bleed off at the brakes when selecting a lesser metered pressure. The brake system itself stopped the airplane very effectively. During crosswind landings, the pilot insured centering the rudder before applying brake pressure to avoid developing a swerve. No significant difficulty occurred during taxi or braking once the pilot became familiar with the system.

During takeoff and initial climb in the clean configuration, the pilot took off in either military or maximum afterburner thrust, using 20-degree flaps with two external wing fuel tanks, and no flaps. The MiG required minimal differential braking for directional control during the initial takeoff roll with the rudder becoming effective at about 50 knots. Full aft control stick raised the nose at approximately 105 KIAS with 2 external wing tanks and approximately 95 KIAS in the clean configuration. Main gear lift-off occurred at approximately 145 and 135 KIAS, respectively. The pilots noted very light longitudinal control force during takeoff rotation and used care to prevent over rotation and subsequent scraping of the tailskid. Landing gear retraction occurred very rapid (approximately 5 seconds) and resulted in negligible trim change, Flap actuation also occurred very rapidly, but resulted in only slight settling at airspeeds above 170 KIAS.

The exploitation team did not use an external source to measure takeoff airspeed. Takeoff distances varied to some extent with airplane attitude at lift off which was generally a function of pilot technique.

During climb performance checks, the exploitation team used both maximum afterburner and nonafterburner thrust with no external stores. The exploitation team performed the maximum afterburner thrust climbs from approximately 10,000 feet to near combat ceiling (500 feet per minute rate of climb of the aircraft), while performing the maximum nonafterburner thrust climbs from approximately 15,000 feet and termination at approximately 40,000 feet. Their conditions for takeoff performance called the gross weight of a clean plane at brake release not exceeding 11,700 pounds, tanks clean, the wing tanks not exceeding 13,200 pounds, and a runway slope of 4 ft/1,000 ft downhill.

Maximum afterburner thrust climbs showed the combat ceiling to be 51,560 feet, which agreed with the predicted data. At 36,000 feet, the exploitation team found the test day rate of climb for maximum afterburner thrust twenty percent less than predicted.

Maximum nonafterburner thrust climbs generally showed less rate of climb than the predicted data. At 36,000 feet, the exploitation team found the test day rate of climb for maximum nonafterburner thrust approximately twenty-two percent less than predicted.

To evaluate level acceleration/energy maneuverability, the exploitation team performed maximum afterburner thrust level accelerations at altitudes throughout the operational envelope of the airplane having no external stores. The exploitation team also performed maximum nonafterburner thrust level accelerations both with no external stores and with two external fuel tanks at several different altitudes. The exploitation team initiated accelerations at approximately 200 KIAS and terminated near the design limit of the airplane. Level acceleration capabilities were good but less than predicted.

While evaluating maneuverability of the MiG-17F in maximum afterburner thrust with no external stores, the exploitation team found the rate of climb potential (specific excess power) good throughout the flight envelope. The best climb speed for maximum afterburner thrust occurred between .78 and .82 Mach number, and the best climb speed in maximum nonafterburner thrust, approximately 305 KCAS. Notably, it required approximately 11.5 percent maximum nonafterburner thrust fuel flow rate than predicted.

For stabilized level flight performance, the exploitation team conducted speed power tests in level flight to obtain the aircraft specific range in nautical air miles traveled per pound of fuel used. To optimum cruise Mach numbers for no external stores and two external fuel tank loadings, the exploitation team compensated for decreasing weight by flying successively higher altitudes

The exploitation team found the engine thrust adequate for this aircraft and satisfactory engine response above 7000 rpm while making afterburner climbs to 51,000 feet without blowout. The exploitation team used afterburner without problems up to 45,000 feet except during one flight during which required the pilot to descend to 38,000 feet to obtain an afterburner relight.

The operating instructions for the aircraft specified no afterburner relight within three minutes after an afterburner shutdown to prevent exceeding afterburner fuel pressure limits. However, test results indicated that these time limits varied with altitude, and the accomplishment of successful relights after one minute with no overpressure. The afterburner refused to light with afterburning operation selected until the engine reached military thrust.

At high altitude with full power (military or afterburner), the exploitation team exceeded engine rpm limit, however the exploitation team did not encounter compressor stalls with rpm above 7000 (maximum was 11,560 rpm), except during rapid throttle movements above 45,000 feet. The MiG-17F encountered compressor stalls at 47,000 feet while advancing the throttle. Particularly impressive was the lack of exhaust smoke from the engine in either military or afterburner range, and no engine vibration encountered with rpm above 7,000.

The maximum test day Mach numbers with no external stores ranged from 0.97 to 0.98 with maximum level flight speed data obtained at the conclusion of maximum afterburner accelerations.

The airplane demonstrated outstanding maneuvering performance, both in maximum attainable g and high g level reached before buffet onset. During level deceleration, the pilot at time zero extended speed brakes and retarded the throttle to idle while aiming for a constant altitude during the maneuver. This reduced the initial velocity by 20 percent within 15-16 seconds and by 50 percent within 37-50 seconds at the test altitudes. During descent, idle power descent performance and test data showed the time to descend from 43,000 to 30,000 feet at approximately 200 KIAS with speed brakes retracted was 7.8 minutes. Descent from 30,000 to 14,000 feet took 8.2 minutes at approximately 190 KIAS with speed brakes retracted, showing the higher speed 200 KIAS a better indicated airspeed to hold for obtaining maximum distance.

MiG-7 Combat flights at Area 51

A year later, and following the MiG-21F technical exploitation, follow-on exploitation of the MiG-17F FRESCO began under the code names HAVE DRILL and HAVE FERRY.

The Aerospace Defense Command and US Navy attack and fighter squadrons under the overall management of the FTD conducted both technical and tactical evaluations of the FRESCO C under simulated combat conditions against the F-4, F-5, F-8, F-100, F-102, F-104, F-105, F-106, A-4, A-6, and A-7 aircraft.

The AFFTC and NATC evaluated the systems of the two MiG-17F planes for performance, stability, and control. The Aeronautical System Division and Research & Technology Division Laboratories evaluated the MiG-17 systems and vulnerability. The Armament Development and Test Center, NWC, Naval Missile Center, and FTD conducted infrared measurements and CIA contractor EG&G Special Projects conducted radar cross-section measurements and provided radar tracking of the MiG-17F for the FTD and Naval Missile Center.

To evaluate combat flights, the exploitation team made the cruise out and back at 20,000 feet at the speed for best range. The exploitation team used their test data from the technical phase of exploitation and calculated two typical combat flights, one calculated for the clean airplane and the other for two external fuel tank loadings. The exploitation team based each flight on a three-minute sea level combat at afterburner thrust and a five-minute military thrust sea level combat. The exploitation team allowed five minutes for warm up and taxi which used 70 liters of fuel. Takeoff and acceleration took one minute and used 62 liters. The exploitation team used the same intercept and return headings for all the climbs and descents. The exploitation team expended 268 pounds of ammunition during the tests. Descent to combat took three minutes and used 53 liters. The exploitation team allowed five percent reserved fuel, 71 liters for the clean airplane and 110 liters for external tanks, which the pilot dropped when empty.

Combat radius for the clean airplane and three-minute afterburner thrust sea level combat covered 115 nautical miles, while the combat radius for the external tank loading covered 215 nautical miles. With external tanks, the combat radius extended 100 nautical miles, approximately 87 percent. The time for intercept for these two flights was 18.6 minutes for the clean airplane flight, while the external tank-loading flight required 36.2 minutes.

The exploitation team evaluated longitudinal speed stability in the cruise (CR) configuration in conjunction with military and maximum afterburner level flight accelerations. The airplane exhibited positive stick-fixed speed stability (increasing TED or airplane nose down (AND) elevator as speed increased) at low speeds. However, stability weakened as speeds increased to .75 Mach and became essentially neutral in the .75 to .85 Mach range. Between .85 and .97 Mach, which was only attainable in level flight with afterburner, mild but unobjectionable transonic trim changes occurred. When decelerating to the stall, speed stability remained positive until approximately 20 knots above the stall. From 20 knots above stall-to-stall, stability decreased to neutral and under some conditions became negative as the plane encountered pitchup. External tanks did not affect the level of stick-fixed stability although addition of tanks caused an AND pitching moment which required approximately three degrees of ANU elevator to counteract.

The evaluation of the longitudinal stick force variation with changes in airspeed was qualitative in nature due to the lack of stick force instrumentation. Due to the nature of the longitudinal control system, which was a mechanical, reversible system, only air loads on the elevator generated stick forces. A lightly positive stick force gradient existed at all subsonic speeds.

The force gradient reversed slightly between .85 and .93 IMN and resembled typical Mach tuck. At Mach numbers above .93, a very sharp increase occurred in positive force stability with an ever increasing gradient out to limit Mach. The longitudinal stick force characteristics during cruise resulted in little need for pitch trimming, and it was easy to maintain a desired speed.

Elevator movement at the start of some accelerations were due to trim changes caused by rapid application of engine thrust and elevator inputs necessary to reduce airplane pitch attitude to that necessary for a constant altitude acceleration.

The exploitation team evaluated maneuvering longitudinal stability during windup turns conducted at approximately 10,000, 20,000, 30,000, and 40,000 feet over a wide range of Mach numbers. At a given trim condition, elevator per g gradients was essentially linear until the onset of buffet, which was in accordance with classical aerodynamic behavior for a mechanical, reversible flight control system.

Test results indicated a sharp reversal in the elevator per g gradient and a pitchup tendency as the plane approached its lift limit at altitudes in excess of 25,000 feet. At airspeeds with a relatively high lift limit g, a moderate buffet preceded the pitchup that normal pilot reaction generally controlled. However, sudden onset of pitchup at low to moderate airspeeds often resulted in a snap stall and/or spin despite pilot anticipation of these characteristics. This pitchup tendency limited high altitude maneuvering flight in a combat environment.

Maximum gradients were at the lowest airspeeds and minimum gradients were in the .80 to .90 Mach region. Above .90 Mach the elevator per g gradient increased rapidly which resulted in the high stick forces above .90 Mach as discussed in a following paragraph.

Maneuvering stick forces ranged from extremely light to extremely heavy. Pilot comment based this on the lack of stick force instrumentation. The exploitation team found the longitudinal stick forces tolerable below .85 Mach and the aerodynamic lift limit or structural limit readily obtainable. At speeds in excess of .85 Mach and particularly above .90 Mach, the stick became extremely rigid and the airplane was stick force limited to g values less than the aerodynamic or structural limit. This was true even when the pilot used two hands on the stick. The high stick forces presented a limiting factor in a combat environment when maneuvering above .85 Mach.

In summary, although the heavy control forces above .85 Mach and the high altitude pitchup tendency just described were tactical limitations, the MiG-17F generally possessed turn capability significantly superior to any US jet fighter airplane. The great majority of tactical engagements against the MiG-17F in SEA occurred in the low altitude regime where the FRESCO C low wing loading and 8.0 g structural limit were best optimized and utilized. The plane's outstanding maneuverability in this area permitted this rather old and simple fighter airplane to remain such a potent threat in the day of sophisticated modern weaponry.

To evaluate rolling performance, the exploitation team measured lateral control with aileron boost ON over a wide range of Mach numbers at 10,000, 20,000, 30,000, and 40,000 feet with the airplane in the clean configuration. The exploitation team conducted limited testing with the airplane loaded with two external tanks, with aileron boost OFF and with partial lateral stick deflections. The clean airplane with aileron boost ON exhibited poor rolling performance of peak roll rates only between 100 and 130 degrees/second. These peak roll rates occurred at .60 Mach at 10,000 feet, and .65 Mach at 20,000 feet, and 80 Mach at 30,000.

Peak roll rates did not vary appreciably with altitude but did decrease rapidly at airspeeds both below and above the Mach number for peak roll rate. Adverse yaw was practically nonexistent at all conditions tested. With the instrumented airplane, maximum roll rates to the right were generally 10 to 20 percent below that to the left although aileron deflections were, if anything, greater to the right.

The addition of two empty external fuel tanks to the airplane did not affect the peak roll rate although initial roll response was slower as witnessed by increased time to bank 90 degrees and decreased bank angle in the first second when compared to-clean airplane data. The exploitation team did not conduct lateral control tests with fuel in the external tanks.

With aileron boost OFF, the exploitation team experienced very low lateral control, stiff ailerons, and high control forces. The exploitation team obtained a maximum roll rate of 38 degrees per second at .64 Mach and 10,000 feet with boost OFF as compared to 107 degrees per second boost ON. Corresponding aileron deflections were 5 degrees boost OFF and 14 degrees boost ON, The partial deflection of ailerons with boost OFF resulted from the inability of the pilot to overcome the very high stick forces.

The maximum aileron deflection during ground control cycles was +16 degrees. The exploitation team obtained 16 degrees deflection at 200 KCAS, but obtained only four to eight degrees deflection at 525 KCAS. Extrapolation of aileron deflection data to higher airspeeds indicated that at the limit airspeed of 572 KCAS the exploitation team obtained little or no aileron deflection. Verifying this, at near the maximum speed limit, the aircraft rolled off, and the pilot had to reduce airspeed to counteract the rolling tendency.

The lateral control forces were high particularly at or near maximum airspeed. The aileron trim switch was inconveniently located below the left canopy rail but not used nor required throughout the aircraft operational envelope as the lateral control system lacked any tendency to oscillate following abrupt stick inputs. Lack of instrumentation for sideslip angle and control forces prevented a complete analysis of sideslip characteristics. Rudder deflection and sideslip increased positive effective dihedral (right aileron-control for right sideslip). Bank angle requirements were between 10 degrees and 30 degrees. Increasing TEU elevator was required for increasing sideslips; however, elevator requirements were not excessive.

In the PA configuration at 190 and 150 KIAS, dihedral, side force and elevator characteristics were similar to the CR configuration. Increased aileron control (3/4 to full stick deflection) was required for full rudder deflection-steady heading sideslips

The plane required approximately 8 to 12 degrees of aileron deflection and 15 to 20 degrees of bank angle at 190 KIAS, and ten to 15 degrees of aileron deflection and bank angle at 150 KIAS. The larger of the values given above apply to speed brakes in and the smaller to speed brakes out. The exploitation team noted no adverse characteristics during sideslips in either the PA or CR configurations.

The exploitation team investigated lateral-directional dynamic stability throughout the Mach-altitude range in the CR configuration and at 150 and 180 knots in the PA configuration. Abrupt rudder pulses or releases from constant heading sideslips established Dutch roll oscillations. The airplane flew lightly damped (fair) over most of the flight envelope and very lightly damped (poor) near .90-.93 Mach number and in the PA configuration. One lateral-directional maneuver at .9 2 Mach number at 40,000 feet continued undamped for 30 cycles and required pilot action to terminate the oscillations. Damping ratios in PA varied from essentially zero with speed brakes IN to .04 with speed brakes OUT. The exploitation team noted no divergent trends in either the PA or CR configuration.

The damped Dutch roll frequency (f) varied from .35 cps at low speeds to about 1.0 cps at high speed and low altitude. The frequency was a clear function of Mach number and dynamic pressure and appeared to be independent of airplane configuration at the lower speed.

Flight test demonstrated that the directional mode (little roll coupling) dominated the Dutch roll oscillations at the higher Mach numbers with the pilot noting a characteristic snaking motion. Reducing Mach number and increasing trim angle of attack at a given altitude caused the lateral-directional oscillations to couple more progressively at low speeds. Rudder breakout forces remained negligible, allowing the pilot to easily induce mild lateral-directional oscillations. Turbulent weather degraded the lateral-directional flying qualities for the gunfire operation at 400 knots between 5,000 and 10,000 feet, making 350 knots a better airspeed. Carrying wing tanks showed no significant effect on the dynamic characteristics.

Stall characteristics in unaccelerated CR and PA configuration generally exhibited adequate warning in the form of airframe buffet and some stick force lightening. Lateral instability in the form of controllable wing rock occurred from 5-8 knots prior to the stall. The stall itself was mild and generally defined as nose drop and simultaneous right wing roll off. The pilot recovered from a stall anywhere during the approach or stall itself by placing the stick forward of neutral. At stall, pitchup became dependent to some degree on the thrust setting and the airplane's eg. High thrust settings resulted in a steep nose attitude at the stall and therefore probably caused a higher degree air elevator blanking and consequently a more pronounced pitchup. PA configuration stalls exhibited essentially the same characteristics except that the pilot encountered no pitchup when using idle power. At no time did the pitchup become severe or uncontrollable.

Moderate to heavy buffet preceded accelerated stalls and lateral instability as the plane stalled. Stick force lightening, and a subsequent pitchup tendency preceding all accelerated stalls became pronounced above 25,000 feet. At the higher altitudes, the stall generally occurred in the form of rapid uncontrollable pitchup or snap roll. The pilot quickly obtained positive recovery by placing the stick forward of neutral, which tempered the stall characteristics to some degree. At lower altitudes, the pilot experienced a less severe pitchup tendency that normal pilot reactions generally prevented entering a stall. Deep stall penetration without promptly initiating recovery caused the plane to rapidly enter a steady state spin as is discussed in the following section.

The exploitation team encountered one fully developed spin during flight test at the end of a windup turn at 230 knots and 38,500 feet. Pitchup started at 12 seconds with the airplane at 3.2

g and 12 degrees per second pitch rate in a right turn. At 13 seconds, the airplane had pitched up and over into a spin to the left. The spin was substantially erect, oscillatory in pitch with a slightly nosedown attitude, and stable with a spin rate of about 120 degrees per second. The yaw rate at 14 seconds pegged at 30 degrees per second nose left, and the indicated velocity began to drop.

Application of full right rudder against the spin for four turns proved ineffective, however popping the stick full forward quickly at 23 seconds accomplished recovery with an altitude loss during the fully developed portion of the spin (approximately four turns) of 3,000 feet.

Factors contributing to this sudden pitchup and spin during maneuvering flight included their operating in a region of the Mach-altitude envelope where low airspeed and high altitude dictated a higher angle of attack from trim, thus reducing the angle of attack margin available for maneuvering. Low pitch damping because of low air density (i.e., low maneuver margin) contributed to pitchup along with a substantial pitch rate developed in the turn and probable overshoot of the maneuver point.

During approach and landing, the airplane demonstrated normal characteristics with the base leg speed approximately 180 KIAS, and the best final approach speed at approximately 145 KIAS with landing gear down, flaps at 60 degrees down, and speed brakes extended. Extending speed brakes improved the lateral-directional dynamic characteristics and provided a higher approach engine rpm, which resulted in quicker engine response in the event of a go-around. Longitudinal trim was insufficient in the power approach configuration; however, stick forces were very light and pilot compensation was easy. The pilot touched down at 125 KIAS, exercising care during crosswind landings to center the rudder pedals prior to braking since displaced rudder pedals resulted in a swerve upon braking.

Comparing flights for quantitative performance evaluation and qualitative handling qualities comparison, the exploitation team flew two FRESCO-Cs together. Airplane #2 performed better in military thrust level accelerations and climbs while airplane #1 performed better in maximum afterburner thrust level accelerations. The handling characteristics of the two were essentially the same though airplane #2 did appeared to roll slightly faster to the right than to the left, whereas airplane #1 did the opposite.

Level acceleration characteristics of the two airplanes occurred with the vehicles starting side by side at 180 knots and accelerating in military thrust to VMAX. Airplane #2 was about 18 knots faster at the greatest separation and 12 knots faster at the terminal airspeed. Both airplanes reached essentially the same terminal speed though airplane #1 accelerated faster in maximum afterburner thrust with a maximum differential of 30 knots.

Comparison climbs to 40,000 feet in military again proved airplane #2 to be the better performer at this power setting with airplane #1 about 3,600 feet or 2 minutes behind in the climb at 35,000 feet.

Not knowing the duty life and fuel control trim for each engine made it impossible to fully explain the difference in performance in military power. The reversal in performance when the airplanes operated in maximum afterburner might have been due to afterburner nozzle area discrepancy or efficiency of afterburner spray nozzles.

During flight control system control harmony evaluations, the exploitation team flew the aircraft without pitch, roll, or yaw stability augmentation. The exploitation team used power-boost for the ailerons with manual control of the rudder and elevator. The exploitation team found that flying the aircraft without aileron boost, the inoperative boost made the ailerons stiff and consequently slow to react above 400 KIAS. The aileron control system presented very light breakout and friction forces. The exploitation team experienced no breakout or friction forces with the elevator or rudder control systems nor was there any free play in the control linkages. Control harmony while airborne was excellent between about 200 and 450 KIAS. At Mach numbers above .90 IMN or airspeeds in excess of 450 KIAS the longitudinal control force was very heavy. At airspeeds below 200 KIAS, the longitudinal control force was very light and care was required to prevent over rotation during landing or takeoff. The exploitation team considered the control stick as too long and too close to the pilot. Most likely the design for the length of the control stick was for the additional leverage required at near limit airspeed. A more comfortable grip would have resulted had the control stick been 2 to 3 inches further forward and 2 to 3 inches lower. Nevertheless, the pilot readily adapted and disregarded the awkward feeling after one or two flights.

The exploitation team located an electrically operated tab for longitudinal trim on the left elevator, and a switch for "beep" operation of the trim tab under the forward left canopy rail that the exploitation team considered unsatisfactorily located and having an unsatisfactorily slow trimming rate. Sufficient trim was available except in the PA configuration where full ANU trim was insufficient. This was acceptable since longitudinal stick forces were very light at low airspeeds.

The left aileron contained an electrically operated tab for lateral trim and an aileron trim switch located on the left console. The pilots found this location poor but acceptable due to the infrequent requirement for lateral trim. Trim switch operation and rate of trim were satisfactory with sufficient lateral trim available over the operational flight envelope for the symmetrically loaded airplane. Directional trim consisted only of a bend tab on the rudder. The exploitation team found no need for any other directional trim system.

During airspeed calibration, the exploitation team determined the position error of the wing boom pitot-static system by stabilized pace and smoke trail acceleration methods finding a very large change in position error in the transonic region.

The primary objectives of the test program were to determine the tactical capabilities of the aircraft and to accomplish a limited performance and stability evaluation. The exploitation team did not conduct specific system tests, relying instead on observations and a qualitative evaluation of system performance as follows:

The exploitation team found that the hydraulic system operated the landing gear, flaps, speed brakes, afterburner nozzle, and aileron boost.

The speed brakes located on the aft fuselage were of interest in that two different switches in the cockpit actuated them. Depressing and releasing a button located on the control column controlled the speed brakes by regulating the hydraulic pressure. The pilot operated a switch located outboard of the throttle to keep the speed brakes open for longer periods, and a button on the control column for rapid modulation of airspeed during maneuvering flight.

The afterburner nozzle actuating system, considering the period of the aircraft's development, was an efficient and easily maintainable system. Three hydraulic actuators positioned an operating ring, which surrounded the nozzle leaves. With the ring at full aft, the nozzle opened to the maximum. When the ring was full forward, the nozzle closed. Exhaust gas pressure opened the nozzle leaves and actuation of the ring closed them. The exploitation team found this system trouble-free throughout the program.

The pneumatic system consisted of two 2 liter (2.1 quarts) main tanks located in the compartment forward of the landing gear struts, a 2 liter tank located at bulkhead 2, and the main landing gear strut cavities. These operated the wheel brakes, emergency lowering of the landing gear and flaps, charging of the guns, the canopy seal, and the windshield deice.

Instead of using conventional toe brakes to activate the wheel brakes, the exploitation team used a brake lever located on the forward part of the control stick. Depressing the lever gave an increasing braking force up to a maximum of 11 kg/cm 2 (151 lb/in .2). For differential braking during turns, the pilot used differential rudder pedal displacement.

The gun charging system permitted inflight gun loading by the pilot to reduce reaction time required to launch an interceptor force. The design eliminated need for ground gun-arming procedures and safety. This also allowed the pilot to clear a gun jam in flight. Several times during the test program, the gun jammed during firing passes. Actuation of the gun charging system cleared the jam and permitted further gun firing. During these firing passes, the pilot did not notice any excessive pipper jitter, excessive cockpit noise or gun gasses in the cockpit. The only problem area was the bright flashes from the gun muzzle blast during firing being disconcerting to the pilot during night gunnery flights.

For emergency operation, the landing gear and flaps actuated by manual handles on either side of the cockpit below the instrument panel. These handles released the gear uplocks. To lock the gear in the down position, the pilot turned valves located on the right console, which directed air pressure to the system. If only a nosegear needed extending, the pilot pulled the right handle for airloads to lock the aft-extending gear in the down position.

The electrically controlled trim surfaces located on the left elevator and left aileron having no rudder trim was conventional though the exploitation team found the switch placement unsatisfactory for a fighter-type aircraft. For example, the elevator trim switch with a white light incorporated in it to show neutral trim was located on the left canopy rail forward of the throttle, and the aileron trim switch was located on the left console. The control stick grip housed beep-type switches that made the trim appeared secondary to gunfire needs and bomb switches.

The engine control system consisted of the throttle for normal engine and afterburner control and a stopcock lever used for starting and stopping the engine. The throttle contained several interesting features. Movement of the throttle forward of the idle (there was no stopcock provision) engaged a midrange detent. Apparently, this stop positioned for normal cruise operations (approximately 10,500 rpm), making the throttle usable in the cruise to full military range. To initiate afterburning the pilot depressed a button on the throttle and moved the throttle forward to engage an afterburning microswitch. The throttle locked into this position.

While the afterburner thrust was not variable, the exploitation team obtained variable thrust by depressing the same button on the throttle to release it from the afterburner locked position and modulating the normal engine thrust by movement of the throttle between the cruise and full military power stops. To shutdown the afterburner, the pilot retarded the throttle aft of the cruise stop. The cockpit contained an afterburner fuel pressure gage and cut-off switch that the pilot shut-off if the afterburner failed to shutdown when retarding the throttle aft of the cruise stop. However, the switch only stopped ignition and reduced fuel pressure. If the throttle was not retarded aft of the cruise stop the afterburner nozzle remained open with a consequently large loss of thrust.

A separate stopcock/starting fuel-metering lever located between the pilot's seat and the lower portion of the left console started and stopped the engine. After initially actuating the ignition and booster switches, the pilot moved the throttle rearward to engage a microswitch, making this operation essentially one-handed. This action connected the starter system with the starter button on the throttle depressed and the RPM monitor. The pilot then moved the stopcock lever down to the START detent and rpm and monitored EGT. Movement of the stopcock lever to full down gave normal engine operation, however too rapid movement of the lever could cause engine over-temperature problems.

The cockpit pressurization and environmental control systems performed adequately despite the lack of automatic features normally associated with these systems. The environmental control initiated the cockpit pressurization valve and automatically regulated it above approximately 7000 feet. The environmental control required excessive movement of a valve to obtain the desired air temperature in the cockpit. Defrost did not adequately prevent canopy/windshield fogging during descent from high altitude.

The exploitation team found the armament and electronic systems designed primarily for the limited day fighter role. The SCAN FIX radar only provided range information (a green light gave an in-range indication) and was effective to one nautical mile. The pilot was required to use the optical sight for azimuth information. One problem area encountered was determining which target the radar locked on in a multi-target environment. The pilot was required to break lock, reacquire the target, and then determine if the new target and the range cue on the sight coincided. The pilot lost valuable time in a fast-moving aerial combat maneuvering situation while performing this task.

The armament consisted of one 37 mm cannon and two 23 mm cannons mounted on a lowerable carriage beneath the cockpit. The gun carriage was equipped with a hoist, which consisted of a winch, four cables, and a system of rollers. Ground personnel used a special wrench to operate the winch system for loading and maintenance. The exploitation team found this to be a convenient method to gain access to a system usually well submerged in an aircraft's fuselage. However, the exploitation team found it difficult to ground clear the guns following a firing mission. A shell casing remained in the chamber of each gun and took an excessively long time to remove. (This difficulty could have resulted from not having the proper armament tools to perform this operation.) The cannons rigidly mounted at two points to prevent recoil. A third supporting attachment fastened the barrels to the fuselage bulkhead by clamps to reduce dispersion.

The reliability of the airplane was outstanding. The HAVE DRILL vehicle accumulated 131-3 hours during 172 flights in 87 elapsed days. The exploitation team flew four to five flights daily, and could had flown more except that the exploitation team limited the flights to daylight hours and required briefing and debriefing time between flights. At no time during the evaluation period was the airplane flown as often as being possible during an operational maximum-effort flying schedule. The exploitation team lost a number of flight days or flyable missions because of bad weather and practically nonexistent logistics because of this being a black program and a foreign-built plane. Considering the secrecy conditions under which the exploitation team conducted the evaluation, the reliability record of the test airplane was not only exceptional, but was a sobering fact.

A weight and longitudinal balance operation conducted prior to the first flight ensured a safe and representative airplane. The exploitation team obtained desired center of gravity limits from the airframe manufacturer's operational curve of gross weight vs. cg for the MiG-17F. Although the MiG-17F (FRESCO D) was somewhat heavier than the MiG-17F (FRESCO C), external geometry differences were slight. The documented operational cg boundaries were 21.1 percent forward and 29.6 percent aft with gear down, including the expenditure of ammunition. The exploitation team did not know the absolute cg envelope based on elevator to pull CL for the forward limit and neutral point for the aft limit, however, the stated operational bounds appeared generous within any such extremes as evident from published aerodynamic data and flight experience.

Sufficient dimensional data were available to locate the leading edge of the mean aerodynamic chord, the centroids of all fuel tanks, and the centroid of a standard weight pilot/chute. The reference datum chosen for all longitudinal moment arms was the center of the nose jack pad, located several inches aft of the inlet plane.

The exploitation team conducted the initial weighing with the airplane level and all instrumentation installed, wing tanks off, canopy closed, gear and flaps down, pilot and chute out, engine and hydraulic oil in, and empty of fuel.

The exploitation team fitted three load cells of a Cox & Stevens portable weighing kit in the wing jack pads and the nose jack pad. However, the dry airplane without pilot and chute balanced aft, and the nose load cell did not seat properly. Therefore, the exploitation team did not fully fuel the airplane, and used a Fairbanks-Morse platform scale under the nose wheel instead of the load cell at the hose jack pad.

Although official flight restrictions were not issued £or the tests reported herein, the exploitation team made every attempt to stay within flight boundaries recommended in publications.

Maximum speed restrictions obtained from Reference Maximum allowable load factors from the same source were the condition with no external stores, with filled external tanks, with empty external tanks, and maximum allowable load factor g. The exploitation team did not test the airplane with other possible loadings such as missiles, rocket pods, or bombs.

In conclusion concerning of their evaluations, the exploitation team found that the MiG-17F demonstrated outstanding lift-limited maneuvering capabilities with high available g and g-level for onset buffet and a low turn radius. Thrust limited turning performance was also good. Longitudinal stick forces increased significantly beyond .85 Mach number and were excessive above .90 Mach. Lateral-directional damping was generally weak and especially poor near .92 Mach Number Lateral-directional tracking for gunfire suffered from turbulent weather or small, inadvertent control disturbances. Roll rate capability was low, only 100-130 degrees per second. Sufficient aileron deflection was not available past 572 KIAS to prevent the airplane rolling off.

Its internal fuel capacity, with no external stores, limited its range and endurance to a 115 nautical mile combat intercept radius. This included a five-minute warm up and taxi, a military climb on heading to 20,000 feet and cruise at that altitude, three minutes combat in afterburner near sea level, a return heading military climb to 20,000 feet, an idle descent and 5 percent fuel reserved . With wing tanks, the radius extended to 215 nautical miles with the wing tanks dropped after depletion.

The exploitation team found the airplane very reliable with exceptional operational availability. The exploitation team easily accomplished four to five flights day after day with an operational record resulting from a deliberate design approach toward airframe, engine and system simplicity and reliability. Attractive features included dependable in-flight gun clearing and charging, a rear view periscope, and a smooth, well-balanced flight control system. Particularly impressive was the lack of engine exhaust smoke at lower altitudes.

The MiG-17F was easy to fly, and the undesirable features, such as spin and accelerated stall, readily handled when encountered. With proper user knowledge of its maneuvering capabilities and limitations, the MiG-17F proved to be a very effective interceptor/air superiority daylight fighter throughout most of the subsonic flight envelope.

Radar Cross Section evaluations

The CIA's EG&G Special Projects contractor exploitation team performed the MiG-17F radar cross-section measurements for the USAF FTD and the Naval Missile Center, Point Mugu, California as part of the technical exploitation. For this phase of the exploitation, the radar cross section measurement system was composed of six basic radar subsystems controlled by one reference bull-gear servo assembly designed by EG&G. The EG&G Special Project Exploitation team operated and monitored the measurement evaluations from a master control facility in a converted two-story former Navy barracks located adjacent to the RATSCAT radar array, using only four of the radar systems in the measurement program. The exploitation team used the VHF, S, and C-Band equipment to gather the reflectivity data, and the Nike X-Band system as the target tracker to generate range information and target bearings. The Special Projects team designed, constructed, and programmed a special computer to interface all systems to ensure data acquisition and performance integrity. The exploitation team slaved the other radars to a reference servo network controlled by target bearing data from the X-Band radar. The exploitation team digitized the range-gated video data received by each radar system and recorded it on tape for offline computer processing. The exploitation team did not use their VHF telemetry system that normally provided target pitch, heading, and roll data.

For the Navy, the exploitation team on the Special Projects Exploitation team conducted X-band cross-section measurements using their CIA Nike X-band radar with the MiG-17F specially modified by their Navy Weapons Center, customer. NWC controlled the flights where the plane made radial passes to obtain nose and tail data, followed by offset passes to obtain broadside data. The plane conducted combat maneuvers to obtain several other aspects. The Special Projects Exploitation team obtained data on several other flights on a non-interference basis, obtaining aspect angle information on these flights by a boresight camera.

Following the cross-section measurements for the Navy, the exploitation team conducted six flights for the US Air Force FTD to obtain VHF, S band, and C-band cross section data. During three of the flights, the exploitation team obtained broadside data and on three, the exploitation team obtained nose and tail data, recording the pulse-to-pulse recordings on digital tape where the exploitation team reduced the digital data to yield median values.

The Special Project exploitation team officially conducted the MiG-17F airborne flights per the specifications and perimeters established by their Navy and Air Force customers. However, for the exploitation team this afforded opportunity to advance RCS technology from that obtained during Mach-3 overhead flights of the CIA's A-12 Project OXCART stealth reconnaissance plane and the previous year's RCS evaluations of the MiG-21 FISHBED. At this point, the Special Projects Exploitation team knew that another RCS program loomed on the horizon - the Project HAVE BLUE stealth prototype that became the F-117 Nighthawk.

In addition to the M-33, mobile Nike X-band fire-control radar, the Area 51 Special Projects data acquisition system consisted of a recording oscillograph and an FM tape recorder that recorded target range, AGC voltage, pulse amplitude, azimuth angle, elevation angle, and timing.

The transmitted signal, coupled to the antenna through a duplexer where a test coupler monitored it while injecting test signals from an X-band signal generator.

A 60 MHz intermediate frequency (IF) amplifier controlled by an AGC-system amplified the received signal with an amplitude modulation 10 Hz upper cutoff frequency. The output of the receiver fed into a boxcar detector, a sample-hold device with return to zero, effectively reducing the bandwidth to 1.5 KHz and amplitude for recording on the oscillograph and tape recorders. The output of the receiver drove the angle and range-tracking systems. A digital-to-analog (D-A) converter converted the digital range information from the range tracker with the linear voltage varying between 0 and 4.5 volts proportional to target range.

The exploitation team received a 2048-yard ambiguity of the range readout that the exploitation team resolved by observing a coarse analog range readout obtained from the analog computer in the form of a dc voltage proportional to range. The exploitation team recorded the range output signals on the oscillograph and tape recorders along with the analog computer's height, ground range, heading and provided inputs for an X, Y, Z plotting board, all synchronized by a timing pulse generator.

For visual observation, optical tracking, and photography, the exploitation team boresighted an optical system consisting of three periscopes with the radar antenna, providing a field-of-view of 6 degrees and a magnification power of eight.

For dynamic cross section of FTD audio frequency measurements collected against the HAVE DRILL vehicle the exploitation team used two NAGRA III tape recorders and three sensing devices for collection. A dynamic microphone for recording sounds in the usual manner, an infra-red radiometer recorded the audio frequency modulations of the infrared radiation from moving targets, and a CW radar recorded modulations of the radar return from moving parts of the target.

The exploitation team used several types of harmonic analysers to separate the mixed signal into its discrete components. The exploitation team associated the frequencies of these components with such mechanical parameters as the number of blades on a rotor stage, or stages, multiplied by the number of revolutions per second of the rotor. The exploitation team analyzed several rigidly coupled rotating units as each radiated a frequency, or frequencies, proportionality the number of blades, or other structural members.

The exploitation team recorded the MiG-17's engine start, idle, taxi, take-off, landing, fixed speed runs at various engine speeds, and engine shut-down using acoustic, infra-red, and radar modulation sensing devices. Bright sunlight conditions proved unfavorable for infrared modulation collection, but did not affect the microphone or radar modulation systems.

Microphone recordings detected a signature characteristic of a centrifugal compressor at all engine speeds from the early stages of the start cycle to late in the shutdown cycle. At idle speed, this signature was a series of harmonically related frequencies with a fundamental of 1450 Hz. FTD estimated the generated 1270 Hz signal by one of the engine-driven accessories, most probably the two variable-stroke plunger type fuel pumps. Melpar, Inc. identified a similar type of fuel pump as the source of such a component in the signature of the J33 engine in the mid 1950s as part of their work in the development of the presented FTD acoustic program.

During this phase of the exploitation, the exploitation team monitored the engine speed under practically all operating conditions by means of suitable microphone recordings. They learned that an engine driven accessory radiated a signal of sufficient strength for detection in the microphone recording, both directly and by its modulation of the engine rotor signal (Beats). Nonetheless, the exploitation team eliminated having exploitable acoustic benefits at distances in excess of one mile.

AFFTC and USN MiG-17F propulsion systems evaluation

AFFTC Propulsion Systems Evaluation Performed By FTD' Naval Air Propulsion Test Center.

For the AFFTC MiG-17F FRESCO C propulsion system evaluation, the exploitation team chose the HAVE DRILL plane serial number 055 to obtain data from the engine test and collect engine installation information. The VK-1F power plant for the FRESCO C aircraft was a 1965 Poland-built afterburning turbojet derivative of the British Rolls-Royce Nene engine and its US version, the J42-P series engine. The J48-P series engine was an improved and uprated development of the basic Nene power plant.

An uncooled single stage axial flow turbine drove the VK-1F engine's single spool double entry centrifugal flow compressor. The engine combustor was of the can type and consisted of nine individual cans. Only combustion chambers number 3 and number 8 had ignitors and had two torch ignitors each. Telescopic sleeves served for balancing the pressure in the combustion chamber and for propagation of the flame during the engine starting procedure.

The VK-lF engine was a continuous flow turbojet engine with nine through flow combustion chambers and incorporated a single stage double entry centrifugal compressor driven by a single stage axial flow turbine.

The afterburner nozzle was a two-position type-full open for afterburning operation and full closed for nonafterburning operation. The afterburner thrust level was modulated by moving the throttle (nozzle remaining full open) and varying the engine RPM. The pilot had to select the maximum afterburning power setting to obtain afterburner ignition. The afterburner consisted principally of a "V" gutter type of flame holder and a cylindrical center pilot unit. Near the "V" gutter were 2k main afterburner fuel injectors, which injected upstream. Two injectors mounted on the rear face of the cylindrical center unit injected the pilot fuel. The V gutter and a round slotted baffle type disc attached to the downstream face of the cylindrical center unit slowed the air down. The slowly moving air then mixed with fuel and the ignition of the mixture in this zone resulted in the pilot flame. Air not utilized by the engine was by-passed around the afterburner and tailpipe section. A small shrouded section of the aft portion of the tailpipe directed by-passed cooling air around the rear portion of the tailpipe, over its external surface and through the annulus of the nozzle leaves.

The two position convergent nozzle consisted of eight overlapped leaves hinged at their forward edge. A hydraulically actuated ring surrounded the leaves and in its most forward positions closed the nozzle leaves for non-afterburning operation. The pilot used a manual override in the cockpit to place the tailpipe nozzle in the full open position during non-afterburner operation with power settings up to and including cruise rated power. This position was for alleviating exhaust gas overtemperature conditions. In its most rearward position, all afterburning power levels, the hydraulically operated ring allowed the nozzle to be in a full open position where exhaust gas pressure forced the nozzle leaves against the nozzle ring.

The engine compressor bleed provided for cockpit pressurization and external auxiliary fuel tank pressurization. The bleed airflow required for these two items was negligible in regard to its effect on engine performance. There were no airframe power extraction requirements on the engine.

The engine and nozzle control system was generally of simple unsophisticated design employing switches and electro hydraulic valves.

The engine start cycle initiated with a starter button. The starter automatically disengaged at a predetermined engine RPM. Fuel and ignition initiated manually. After light off, the engine accelerated at its own rate to idle. Limiting the pilot's advancement of the throttle from idle to military power at less than .1 to 2 seconds prevented exceeding the engine operating-temperature limits.

Both the main engine and the afterburner fuel pumps were of the wobble plate variable stroke plunger type (similar in design to the Lucas plunger pump). The accessory drive gearbox drove both pumps by gearing off the front portion of the mainshaft of the engine. The starting (priming) fuel pump was of the vane (blade) type. The RPM of the pumps determined the quantity of fuel delivered to the starting fuel nozzles and the amount of fuel bypassed through a reducing valve. The amount of fuel by-passed was dependent on the pressure of a spring. The pressure of this spring was externally set. The engine fuel control senses throttle position, compressor inlet pressure, and combustion chamber inlet pressure.

During steady state engine operation, a flyball governor held engine speed constant. The flyball governor operated through a servomechanism to alter the engine fuel pumps delivery rate. In acceleration mode, scheduling the engine fuel flow was a function of combustion chamber inlet pressure, biased however, by compressor inlet pressure. The fuel control had no limiting functions such as primary burner overpressure or exhaust gas overtemperature overrides.

The afterburner fuel dump delivery was proportional to engine RPM and fuel pump wobble plate position. The afterburner fuel control metered fuel as a function of compressor face total pressure and engine RPM. The afterburner fuel distributor provided the proper fuel pressure at the outlet from the fuel pump for all operational flight altitudes and velocities.

The aircraft power plant air induction system for the FRESCO C aircraft was of the rounded lip normal shock fixed geometry type. An inlet of this nature provided satisfactory performance up to approximately Mach 1.6, well in excess of the plane's normal flight velocities encountered. The inlet was bifurcated and there were two horizontal flow dividers on each side of the subsonic diffuser. The starboard and port inlet duct exits were mirror images with the capture area of the inlet and the inlet throat area giving an internal contraction ratio of 0.79.

The VK-1F engine model specification indicated a guaranteed life for the engine of 100 hours; however, other information stated an increase to a standard of 200 hours, and in actual practice, 250 hours frequently elapsing before engine replacement. The exploitation team believed that the engine tested was equipped with original factory equipment. The exploitation team accomplished no engine trim before, or immediately after the HAVE DRILL test program proper.

Operational test exploitation of the MiG-17F

Lt Col Wendell H. Shawler, chief of the 6512th Test Squadron's Special Projects Branch at Edwards Air Force Base, served as project manager. VX-4 test pilots LCDR Foster S. "Tooter" Teague and LCDR Ronald F. "Mugs" McKeown drew up a test plan. Fred Cuthill made the first functional check flight in the HAVE DRILL MiG-17F (YF-113A) on February 17, 1969, the first of 172 technical and tactical evaluations sorties that concluded in the middle of May.

The second MiG-17, code-named HAVE FERRY and designated YF-114C was delivered on March 12, 1969, and completed its first functional check flight on April 9, 1969.

The two MiG-17s flew against numerous US aircraft types during the tactical evaluations. In slightly over a month, the test exploitation team flew 52 sorties in the YF-114C, including dual flights with the YF-113.

The operational test exploitation program consisted of approximately 130 flying hours (from brake release on take-off to five minutes after landing), 152 test flights, and several ground runs. By the end of the entire program, including flights after the test program the engine had acquired 209 flights and 155 total hours.

During the test program, the exploitation team made approximately 260 starts on the engine, encountering only three unsuccessful starts during the program except for 20 to 30 attempts to get the starter to engage on 1 March 1969. The exploitation team attributed the 1 March starting problem to a faulty microswitch while attributing a malfunctioning circuit breaker as cause of the other three unsuccessful starts.

Discrepancies with the engine related systems during the test were as follows:

On 21 March 1969, replaced the cracked and distorted afterburner tailcone with the afterburner assembly from another VK-1F aircraft. A flow check of the removed afterburner fuel nozzles indicated low delivery in several of them.

On 17 April, the afterburner failed to light off at 17,000 feet, but accomplished a successful light off at 10,000 feet.

During an engine ground test on 19 April 1969, the afterburner failed to light off on four successive occasions. The exploitation team cancelled further attempts for light off and found the afterburner igniter misaligned. This could have been due partially to overtemperature expansion during an earlier portion of the ground test.

On 22 April 1969, the maintainers removed, repaired, and installed a malfunctioning circuit breaker. At the time of disassembly, the maintainers found a hole in the aircraft aluminum liner near the afterburner light off zone caused by the liner melting during local overtemperature conditions during the 19 April ground run. Aluminum had splattered on the engine compressor inlet rear face.

On one occasion, the engine nozzle failed to open for afterburner operation. The exploitation team found that an electrical solenoid, mounted on an internal portion of the aft section of the fuselage, was not functioning. This solenoid ports hydraulic fluid into the hydraulic pistons which connected by rods to the nozzle ring.

There were no occurrences of engine oil or fuel pump malfunctions during the test.

There were no occurrences of engine surge or excessive engine vibration during either the sea level static steady state or transient engine test; however, the pilots noted some indications of slight surge during flight.

During the flight test portion of the program engine data retrieved was limited by available space for test recording equipment in the aircraft and by AFFTC requirements. The engine data recorded on an oscillograph was limited to fuel flow and that recorded on the photo panel was limited to fuel temperature and RPM. When required, the pilot monitored steady state values of exhaust gas temperature, and RPM.

The exploitation team conducted two engine ground runs during the test. On the first test, the exploitation team recorded steady state values of compressor inlet pressure and temperature, compressor discharge pressure, turbine discharge pressure and temperature, and exhaust nozzle temperature and pressure. The exploitation team recorded the fuel flow on an oscillograph, RPM and fuel temperature on a photo panel and a thrust indicating device reading thrust measured by load cells. The second ground test run divided into two parts. The Air Force conducted a steady state engine test and the Navy conducted a transient engine test aided by the Air Force. Data acquired during the second steady state ground test run was as outlined for the first ground run, with the addition of turbine discharge pressure. The exploitation team presented only the data generated from the second steady stage engine test run, finding no apparent significant deviation of engine performance during the two tests. Data taken during the second test also included thrust, a parameter that had anomalies in measurement on the first test. For comparison, the exploitation team tabulated only the data obtained along with published Soviet uninstalled data.

Based on the data acquired during the HAVE DRILL test, the exploitation team concluded the engine performance being close to that of a new factory trimmed engine even though the engine had high time. During the test, the engine nozzle area was at specification value in both dry and afterburner configuration. The maximum physical RPM at which the exploitation team ran the engine was on the average 1.12 percent below specification value, which resulted in a reduction in thrust output and fuel flow. Exhaust gas temperature, thrust, and fuel flow after adjustment with the proper correction factors were all within the manufacturers prescribed operating limits.

Soviet specification engine thrust and corrected engine thrust test values received during the test were in excellent agreement,

The nozzle pressure at military power was in exact agreement with the Soviet published data. Nozzle pressure was a parameter in which the exploitation team placed a high degree of credence as to proper engine trim and operation.

The VK-1F engine acceleration times were slow in comparison to Free World operational engines of the same type and vintage. They were however, within the maximum times published in the Soviet literature. Acceleration times were somewhat hindered during the test due to adverse effects of installation and altitude pressure losses.

Engine reliability throughout the test was good; however, the pilots experienced some difficulty with engine controls.

Naval Air Propulsion Test Center VK-1F engine static transient ground test

Consistent with the general objectives of Project HAVE DRILL, the Naval Air Systems Command assigned the Naval Air Propulsion Test Center (NAPTC) a task of evaluating transient characteristics of the VK-lF engine.

The Navy measured compressor discharge static pressure using a transducer tied into the fuel control sensing line. They measured the total fuel flow and afterburner fuel flow by turbine type flowmeters, which required AAFTC modifying the engine plumbing to facilitate these flowmeters. For the ground transient tests, AAFTC disconnected the recording equipment to record the fuel flowmeter and engine speed signals using NAPTC equipment.

They recorded engine speed by measuring the frequency output of the engine tach generator.

To measure the turbine discharge pressure and temperature, NAPTC-fabricated four combination pressure-temperature probes to replace the standard exhaust gas temperature (EGT) probes. They connected the average output of three of the EGT thermocouples to an Air Force Howell temperature indicator installed in the cockpit. The remaining thermocouple output inputted to transient recording equipment, and one total pressure-integrating probe connected to a transducer.

AFFTC and the Special Projects exploitation team recorded all parameters on a 10-channel oscillograph, and total fuel flow and engine speed cross-plots on an x-y plotter in the Special Projects facility.

AFFTC flew three runs with the MiG-17F and found the VK-lF engine somewhat poorer than the J48-P series engine, both engines developed in approximately the same 1947 to 1952 period.

Preflight phase (HAVE FERRY)

The HAVE FERRY aircraft arrived at the site on 12 March 1969. Assembly began on 20 March 1969 and completed in 20 days. During assembly, examined the aircraft for discrepancies followed by a 50-hour inspection where the team noted, repaired, adjusted, and operationally checked the following discrepancies:
- Right main tire worn beyond limits
- Right wing fence bent
- Leads on starting sequence control box loose
- Bottom nozzle actuator leaking
- Hole burned thru aft section shroud left side, 9 o'clock position and also cracked.
- Aft section shroud cracked at 6 o'clock position
- Rivets loose both sides, bottom row, forward fuel cell (8) Left and right struts (Main gear) would not hold air or hydraulic fluid.

During the reassembly and 50-hour inspection, inspected, adjusted repaired and operational checked all systems. Also during this time, the exploitation team installed a UHF antenna, wiring, mounting brackets, radio set, and installed a tape recorder in the cockpit

Flight phase (HAVE FERRY)

The exploitation team flew the functional check flight for the HAVE FERRY MiG-17F on 9 April 1969. The following was a summary of delayed discrepancies that generated during the project.
- IFF gear removed from nose.
- IFF control box removed from cockpit.
- Ballast installed in nose compartment. (78 lbs.)
- Maintainability (HAVE FERRY)

Manhours expended for maintenance on the HAVE FERRY MiG-17

- 56.5 man hours for discrepancies
- 32.5 Total Man Hours for maintenance 26,0 man hours for postflights and preflights (2) 37
- 7 Total Flying hours
- 52 Total sorties
- 2.18 man hours per flying hour
- 1.5 man hours per sortie
- 30 maintenance discrepancies.
- 20 Flying days
- 2.6 Sorties per flying day

HAVE DRILL used this aircraft primarily as the backup aircraft for the HAVE DRILL; however, it flew 25 dual tactical flights.

Scope of the HAVE FERRY evaluation

SCOPE OF THE TEST: The scope of the evaluation of the FRESCO C included the

following items against the F-4, F-5, F-100, and F-105:
- Determination of:
  - Most suitable offensive maneuvers.
  - Moat suitable defensive maneuvers. ;
  - Most suitable flight tactics.
  - F-4 radar detection data to include:
  - Comparative-analysis of APQ-120 and APQ-100/I09 performance.
  - Limited study of LegTee and 180-degree aspects.
  - Altitude effects
- Verification and validation of:
  - Tactical doctrine.
  - Energy maneuverability comparative analysis.
  - Operational restrictions and limitations of the MiG-17-
- Preparation of a briefing and/or motion picture to adequately cover the findings of the test.

CHAPTER 6- Tactical evaluation of the MiG-17F at Area 51

The HAVE DRILL and HAVE FERRY maneuvers

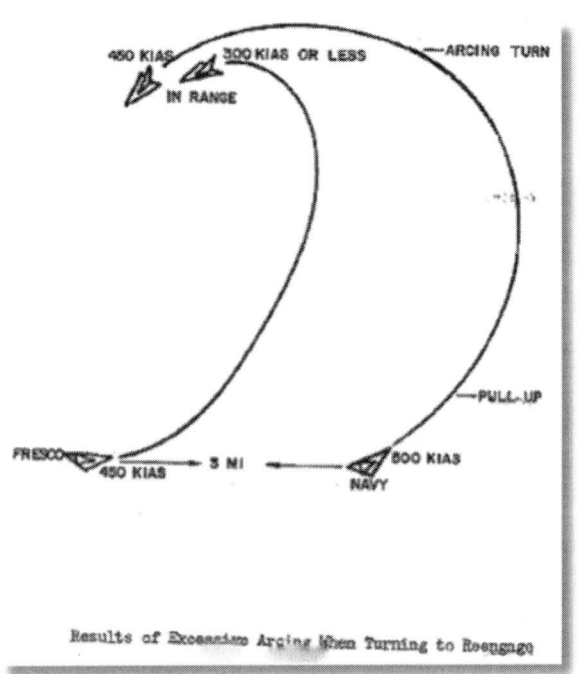

Results of Excessive Arcing When Turning to Reengage

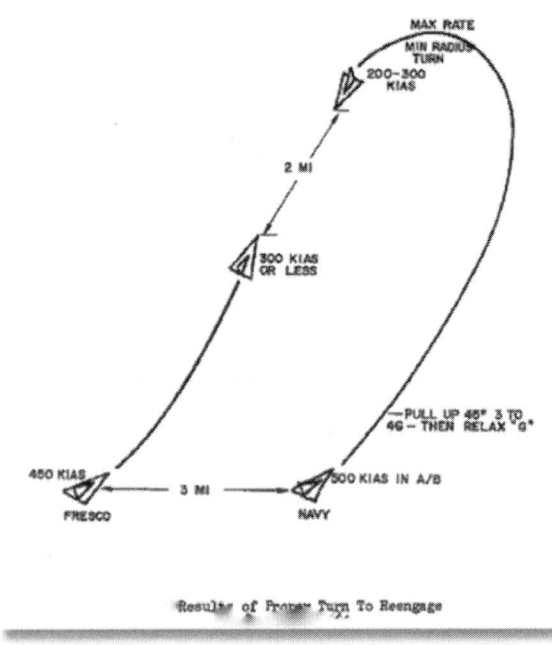

Results of Proper Turn To Reengage

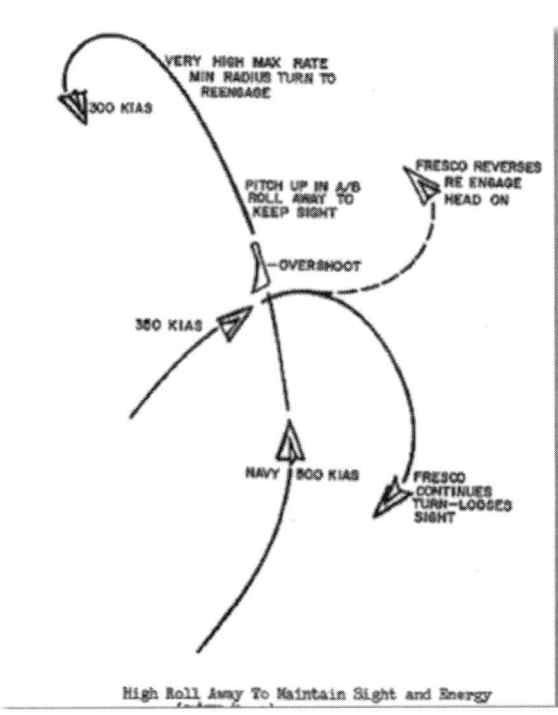

High Roll Away To Maintain Sight and Energy

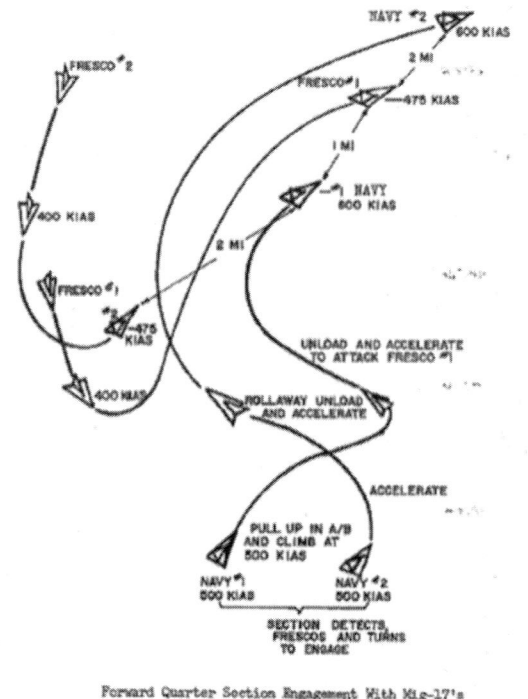

Forward Quarter Section Engagement With Mig-17's

F-4 PHANTOM II vs. MiG-17F FRESCO C

The exploitation team flew twenty-six F-4 flights against the MiG-17F with the F-4, in most cases, configured with a full complement of air-to-air weapons to enhance the realism of the evaluation.

Acceleration comparisons conducted were one of three types: military power for both aircraft, maximum power for both aircraft, and maximum power for the FRESCO C while the F-4 maintained military power, both conducted in level flight and in an unloaded condition. At military power, from 250 KIAS until the MiG-17F reached 400 KIAS, the F-4 out-accelerated the FRESCO C by roughly 50 knots and 5,000 feet, a difference more pronounced at maximum power, approximating 100 knots, and some 8,000-10,000 feet. When the FRESCO C performed at maximum power, it displayed an acceleration potential somewhat better than the F-4 at military power. The majority of the test runs occurred at 15|000 feet, although those that occurred at 30,000 feet did not expose any significant differences. It was worth noting that on flight 1 and again on flight 49, the FRESCO C initially responded better than the F-4, and the F-4 had to accelerate past approximately 270 KIAS to gain an advantage.

Zoom comparisons were performed on flights 1, 2, and 19, to analyze the comparative zoom potential of the FRESCO C with both aircraft operating at maximum power, and with the FRESCO C at maximum power vs. military power in the F-4. With both aircraft at maximum power, utilizing the zoom profile resulted in an altitude advantage of some 4,000 feet and an airspeed advantage of 40 knots for the F-4 flight 3. When the FRESCO C performed the maneuver at maximum power and the F-4 performed the maneuver at military power, the FRESCO C enjoyed an 800-foot and 50-knot advantage.

Turn comparisons conducted on flights 1, 2, and 49, the exploitation team initiated comparisons at Mach .9, with two performed at 15,000 feet MSL and two at 30,000 feet MSL. It was determined that, at those airspeeds, the F-4 turning capability was very near that of the FRESCO C if the FRESCO C did not attempt to decelerate. The exploitation team also learned at a very early point that at airspeeds approximately 350 KIAS, the FRESCO C had a turning capability far superior to that of the F-4. Consequently, the exploitation team expended no effort to further ascertain that fact.

The poor turning performance of the aircraft at high-calibrated airspeeds on previous flights precipitated the roll rate comparisons performed on flight 10. The comparisons occurred at 350 and 450 KIAS, 15,000 feet MSL. At 350 KIAS, the F-4 roll rate was twenty-five percent better (360 degrees -Vs. 270 degrees) than the FRESCO C, and at 450 KIAS, the F-4 roll rate exceeded twice that of the FRESCO C (360 degrees vs. 160 degrees).

A single speed brake deceleration comparison occurred where the exploitation team initiated comparison at 3.5 KIAS from a stabilized line abreast position, 15,000 feet USL. The F-4 had 330 KIAS when the MiG-17F terminated the maneuver with 300 KIAS. The exploitation team concluded that the FRESCO C could decelerate with speedbrakes roughly twice as effectively as could the F-4.

Fresco C and F-4's

The exploitation team attempted radar detection samplings on as many flights as possible, on the philosophy that a small sampling of radar data being inconclusive. Fifty-six recorded intercepts took place at various points throughout the evaluation. The exploitation team completed thirty intercepts utilizing F-4C/D radars (APQ-100/109) and accomplished the remainder with F-4Es (APQ-120).

Within the definition of the data collected, the exploitation team found no significant difference between the performance of the APQ-100/109 and the APQ-120 in a clutter free environment (approximately 22 nautical miles average at 25,000 feet). However, as the tabulated data clearly indicated, at 15,000 feet USL, which averaged 9,000-10,000 feet AGL, both systems declined in detection percentage (F-4G/B, 67% F-4E, 44%).

This fact arose due to the ground clutter problem generally encountered by pulse radars, particularly when attempting intercepts over mountainous or irregular terrain. The lower detection percentage of the F-4E could be attributable to the larger vertical beam width produced by the F-4E radar antenna. A radar with a large beam width would, as could be borne out geometrically, encounter ground clutter problems at a higher altitude than would a radar with a smaller beam width. At 10,000 feet, the F-4 detection probability 12.5% faltered badly. Insufficient parallel data existed for the F-4C/D to draw a similar conclusion, but the trend indicated at 15,000 feet should have continued to decline at 10,000 feet. Taken as a whole, the average detection range for all passes on which detection occurred was 17.7 nautical miles. The significant point gleaned from this discussion was that the performance of the F-4 radar at low altitude was unsatisfactory, and that was precisely where the FRESCO C chooses to operate in Southeast Asia.

RAW data collection occurred on Emissions 32-34. When the range only radar of the FRESCO C operated, it always produced warning indications on the APR-25/26 or ER-14. The radar operated at a frequency of approximately 2,800 MHZ, which was near that of the gun laying radar used by communist-bloc nations in Southeast Asia.

Therefore, on the APR-25/26, the AAA/AI light accompanied a strobe on the cathode ray tube. Accordingly, it became extremely difficult to differentiate between AAA radar and the FRESCO C range only radar in a multi-signal environment without an analysis of APR-25/26 strobe lengths. Such an analysis of a strobe in the rear hemisphere could disclose whether or not closure existed by the eventual behavior of the length of the strobe, and could, thereby, served to alert the crew of the possible presence of a FRESCO C threat.

To summarize, the range only radar of the FRESCO C would illuminate the AM/AI light of the APR-25/26, and aircrews should carefully analyze such an indication before assuming an AAA radar lies at its source.

Offensive and defensive maneuvers by the F-4 against the FRESCO C were evaluated on flights 1-4, 10, 11, 16-21, 44, and 46-49.-These maneuvers were thoroughly briefed and tightly controlled to less than 360 degrees of turn. Offensive maneuvers included high/low speed yo-yos, lag pursuit attacks, and maneuvers in the vertical plane. Defensive maneuvers included hard/break turns, slicing turns, climbing turns, level, and unloaded accelerations, high g barrel rolls over, unloaded rolls, and preplanned high-calibrated airspeed reversals.

In a one-on-one situation and with mutual awareness, the F-4 cannot turn with the FRESCO C unless both aircraft were operating at high-calibrated airspeeds as the maneuvers during flights 2, 3, 11, 19, 44, and 46 pointed out.

The exploitation team judiciously employed yo-yos and lag pursuit attacks because their continued application tended to permit the engagement to progress into low speed, high angle of attack maneuvering, which was very much in favor of the FRESCO.

Climbing turns were marginally effective in permitting the F-4 to maintain an offensive potential and, the MiG-17 demonstrated a better vertical capability than was anticipated when if flew in afterburner power, flights 4, 10, 11, and 18.

Defensively, the F-4 could not generate a significant overshoot by the MiG-17 through hard/break turns into the attack. A slicing defensive turn was more effective than a level defensive turn insofar as it permitted the F-4 to retain energy while testing the ability of the FRESCO C pilot more severely than a level turn, but it was still not a decisive maneuver.

The acceleration potential of the F-4 successfully producing separation in a defensive situation flights 3, 11, 17, and 18, and could had prevented any effective firing passed by the FRESCO C if detection occurred early enough. A lethal position forced the FRESCO C to conduct some sort of last-ditch maneuver to spoil its tracking solution before attempting an accelerated separation. Of such maneuvers attempted, the unloaded roll was the most effective in destroying a tracking solution. A high g barrel roll produced the desired initial results, but diminished energy rapidly and left the F-4 behind the acceleration curve, unable to effect rapid separation. As an added precaution, once acceleration had produced airspeeds that forced the FRESCO C above 450 KIAS, rapid reversals by the F-4 at random intervals denied the FRESCO C lethal position as range opened.

Obviously, any escape maneuver that forces the FRESCO C into a high calibrated airspeed situation would be to the advantage of the F-4 since the maneuverability of the FRESCO C was inhibited in that flight regime. Permitting unloaded rolls and/or unloaded accelerations to develop into steep dive angles while the FRESCO C pilot concerned with recovery of his aircraft beyond other considerations turned the advantage to the F-4.

To summarize, in a one-on-one situation, the F-4 crew should endeavor to arrive at an initial offensive position by all means available. From that position, preferably in the rear hemisphere, weapons employment should observe the philosophy of increasing firepower with decreasing range, or missile attack followed with a gun attack if a gun was aboard. If the FRESCO C was not destroyed, the F-4 should not become involved in a low airspeed, high angle of attack turning battle with that aircraft, but should effect an immediate maximum power level or unloaded acceleration as soon as gun tracking was no longer possible. The pilot could reattack if such a move was commensurate with flight objectives, by accelerating to Mach 1.2 or faster and executing a low (2-3) g vertical zoom while keeping the enemy in sight. Separation required to permit a turn-back should approximate three ITM. The retention of visual acquisition at that range would require the pilot's utmost concentration. The pilot could expect subsequent attacks being head-on passes unless the FRESCO C pilot elected to fall off before the capability of his aircraft so dictates.

Defensively, outside the gun range of the FRESCO C, maximum power acceleration could defeat a gun attack. Inside gun range, at 1,500 to 2,000 feet, the F-4 had to execute a last ditch type of escape maneuver simultaneously with a maximum power acceleration in order to insure survival. Such a maneuver should progress, if possible, into a steep dive angle accompanied with random reversals at high-calibrated airspeeds to deny the FRESCO C a tracking capability during separation. Again, the pilot could reattack if commensurate with flight objectives.

During flights 16, 17, 19, 32, and 3, the exploitation team evaluated F-4 flight tactics employed against the FRESCO C with the F-4 enjoying numerical superiority. Sound enemy tactics combined with competent FRESCO C pilots presented a formidable problem to F-4 crews when the enemy had numerical parity or a numerical advantage. The strategy employed against the FRESCO C on these flight emphasized split plane maneuvering and mutual support between elements. Upon threat detection, the elements split in the vertical plane to force a decision on the MiG-17. Once the enemy committed, the superior acceleration of the F-4 came into play. The F-4s subsequently sandwiched the FRESCO C while staying out of danger individually, and controlled the engagement until obtaining the desired result. The FRESCO C could not disengage and separate under these circumstances without the F-4 firing upon it.

To summarize, the F-4 could defeat the FRESCO C under conditions of numerical superiority by utilizing split plane maneuvering to achieve a sandwich, and complementing that tactic with mutual support. Mutual support was extremely critical. Crews had to not only keep constant pressure on the enemy, but had to keep each other apprised of the enemy's position and actions. Elements could not separate where they cannot visually clear one another, especially in consideration of the fact that FRESCO Cs rarely performed singly. The pilot had to maintain a constant vigil in anticipation of the presence of a yet undetected threat. In all cases, the pilot employed high-calibrated airspeeds during tactical operations to complicate the attack problem for the FRESCO C during all phases of flight when a FRESCO C threat exists.

The following conclusions pertained to an F-4 aircraft configured with four semi-submerged inert AIM-7 missiles and four externally carried AIM-9 missiles and the FRESCO C clean.

- With both aircraft operating at military power, the F-4 had an acceleration capability far superior to that of the FRESCO C above 250 KIAS.
- With both aircraft operating at maximum power, the F-4 had an acceleration capability far superior to that of the FRESCO C.
- With the FRESCO C operating at maximum power and the F 4 operating at military power, the FRESCO C acceleration capability was slightly superior to that of the F-4.
- With both aircraft operating at military power, the zoom capability of the F-4 was superior to that of the FRESCO C.
- With both aircraft operating at maximum power, the zoom capability of the F-4 was superior to that of the FRESCO C.
- The zoom capability of the FRESCO C at maximum power was slightly superior to the zoom capability of the F-4 at military power.
- At 350 KIAS and below, the turning capability of the F-4 was inferior to that of the FRESCO C, markedly so at airspeeds below 250 KIAS.
- At 450 KIAS and above, the F-4's turning capability compared to that of the FRESCO C, but the F-4 bled airspeed much more rapidly. If the FRESCO C utilized speed brakes to decelerate while turning, it retained a turning capability superior to that of the F-4.
- With initial power settings closely matched and speed brakes employed, the FRESCO C decelerated more rapidly than the F-4. An idle power deceleration without an accompanying utilization of speed brakes permitted the F-4 to decelerate at a rate comparable to the FRESCO C.
- The F-4 roll rate with full-scale stick deflection was superior to that of the FRESCO C, particularly at high-calibrated airspeeds.
- The F-4 could best defeat the FRESCO C in an aerial engagement by utilizing hit and separate maneuvers,
- If a reattack was commensurate with flight objectives, the F-4 had to follow a separation maneuver with a high Mach, low g vertical zoom to regain an offensive potential,
- Defensively, the F-4 could readily defeat an attack by the FRESCO C by accelerating to

supersonic speed. If, however, the FRESCO C was in such a position that an acceleration alone failed to immediately produce safe separation, the E-4 employed a last ditch maneuver (see test results and discussion) to defeat the lethal position of the FRESCO C before attempting to separate. If possible, separation had to utilize a steep dive angle,

• Within the definition of the data collected on this evaluation, there was no significant difference between the FRESCO C detection potential of the APQ-100/109 and APQ-120 radars in a clutter-free environment.

• All F-4 radars encounter ground clutter difficulties at altitudes of 10,000 feet AGL and lower. The FRESCO C operating at or below 10,000 feet AGL substantially reduced the effectiveness of the F-4 air-to-air radar and radar missile system.

• Exploitation results concluded that following an initial confrontation with the FRESCO C and a subsequent separation maneuver, the USAF and USN had to apprise F-4 tactical aircrews of the inadvisability of attempting a reattack in a one-on-one situation unless obtaining sufficient separation (2.5 to 3 miles) while maintaining visual contact with the threat. NOTE: If visual contact was lost, it would require the utmost attention of the aircrew to reacquire the target visually and the crew could have to resort to radar acquisition.

• Keep F-4 ingress speeds sufficiently high to permit rapid acceleration in order to defeat a FRESCO C rear hemisphere gun attack.

F-100 vs. MiG-17F FRESCO C

Six test flights (8, 9, 15, 24, 25, and 53) evaluated the F-100 against the FRESCO C, with only F-l00Bs in a clean configuration.

Acceleration comparisons performed found that the F-100 had to accelerate above 300 KIAS before it achieved a degree of superiority in military power. At maximum power, the F-100 was clearly superior to the FRESCO. If the FRESCO C accelerates at maximum power and the F-100 accelerates at military power- the MiG-17 was slightly, but not significantly, superior.

Turn comparisons occurred on all F-100 flights, save number 15. The results of those comparisons varied with the power settings used and initial airspeeds. The F-100 turned faster than the FRESCO C at 450 KIAS or higher, but bled airspeed more rapidly than did the FRESCO C, which accelerated in a maximum performance turn at maximum power. As airspeeds reduced to 350 KIAS, the two aircraft executed a 180-degree turn in a comparable time frame with the F-100 slightly better, as at higher airspeeds, though the FRESCO C retained more airspeed. Although no turn comparisons occurred at slower speeds than 350 KIAS, it was apparent during the offensive and defensive maneuvers that a transition occurs between 350 KIAS and 300 KIAS, below which speed the FRESCO C enjoyed a dominant turn capability.

Roll rate comparisons occurred on flights 8 and 15, and those comparisons disclosed that the F-100 roll rate at 350 KIAS was one-third better than the MiG-17. At 400 KIAS, the F-100 roll rate was twice that of the FRESCO C, and at 450 KIAS, the F-100 rolled level before the FRESCO C had reached an inverted position (360 degrees of roll vs. 160 degrees of roll).

Zoom comparisons performed on flights 15 and 2 demonstrated the F-100s slightly superior zoom potential at maximum power (30 knots and 500-800 feet). At military power, the two aircraft were comparable in a zoom, with a slight edge belonging to the F-100.

Comparative speed brake effectiveness analyzed on flight 8. Started from an initial 390 KIAS in a stabilized line abreast position where both aircraft deployed the speed brakes and slowed to 300 KIAS. With the MiG-17F at 300 KIAS, the F-100 still had 340 KIAS and was more than 1,000 feet in front of the FRESCO C.

The exploitation team employed offensive and defensive maneuvers against the FRESCO C on all F-100 flights. Offensively, the MiG could force the F-100 into an overshoot, regardless of his plan of attack. Defensively, the F-100 effectively utilized afterburner accelerations to depart a vulnerable position. Afterburner acceleration by the F-100, however, could not immediately produce safe separation if being subjected to gunfire.

It followed that some sort of last-ditch escape maneuver would be required in order to prevent tracking while achieving separation. Unloaded reversals at high-calibrated airspeeds during flight 24 were very effective in producing that result. In summary, the F-100 should pursue an attack on the FRESCO C until tracking was no longer possible, maintaining a high attack airspeed. Defensively, the F-100 employed afterburner accelerations to defeat FRESCO C gun attacks. If detection occurred after the FRESCO C had opened fire, the F-100 should maneuver to destroy a tracking solution while accelerating with an unloaded acceleration complemented by rapid reversals at random intervals. If altitude permitted, an escape maneuver should progress toward a steep dive to force the FRESCO C into a high-speed flight if he chooses to follow the maneuver with reattack decisions based on tactical considerations. Generally, a reattack by the F-100 being risky if a competent FRESCO C pilot elected to pursue.

Flight tactics that might be employed against the FRESCO C when the F-100 enjoyed numerical superiority were not evaluated; however, it was apparent that the F-100 effectively employed split plane maneuvering and mutual support to defeat the FRESCO C in a two-on-one situation.

The following conclusions pertain to findings derived from competition between a clean F-100 and a clean FRESCO C.

With both aircraft operating at military power, the F-100 possessed an acceleration capability that slightly exceeded that of the FRESCO. Conducting an acceleration comparison under unloaded conditions showed the two aircraft having comparable acceleration potential.

The acceleration capability of the F-100 exceeded that of the FRESCO C with both aircraft operating at maximum power.

The zoom capability of the F-100 was comparable to that of the FRESCO C, whether conducted at maximum power or military power.

At 350 KIAS or below, the FRESCO C had a turning capability superior to that of the F-100, and below 250 KIAS, the FRESCO C was far superior.

At airspeeds approaching 450 KIAS, the turning capability of the F-100 was comparable to that of the FRESCO C, and perhaps slightly better; however, the FRESCO C lost far less speed in the turn. The FRESCO C regained turning superiority by utilizing a speed brake deceleration in the turn.

At comparable power settings, the FRESCO C decelerated more rapidly with speed brakes than did the F-100.

The roll rate of the F-100 was superior to that of the FRESCO C.

From a position of advantage, the F-100 employed a hit and-run maneuver to defeat the FRESCO C.

When defensive, the F-100 negated an attack by the FRESCO C through the execution of acceleration to supersonic speed, but with a lesser degree of facility than did the F-4 or F-100. Failure of acceleration alone negating an attack by the FRESCO C, required a last ditch maneuver when possible, progressing to a steep dive.

Exploitation results concluded that, following an initial FRESCO C confrontation and subsequent separation, crews avoid attempting to position for reattack in a one-on-one situation, unless they enjoy a tactical advantage.

F-5A vs. MiG-17F FRESCO C

Missions 22 and 23 evaluated the F-5A in a clean configuration in the presence of a FRESCO C for acceleration comparisons. At military power, originating at 250 KIAS, the F-5A accelerated to 400 KIAS and was some 3,000 feet in front of the MiG-17, which had 360 KIAS at that time, when the same comparison occurred in unloaded flight, the F-5A did not obtain a significant advantage until approximately 350 KIAS. Maximum power illustrated a distinct advantage for the F-5, which accelerated from 300 to 500 KIAS while the MiG-17F accelerated from 300 to 410 KIAS.

Zoom comparisons occurred on both flights, one each at maximum and military power. At military power, the F-5 zoomed 1,800 feet above the FRESCO C, which terminated 30 knots faster. At maximum power, the F-5 zoomed 2,500 feet above the FRESCO C and was 80 knots faster at termination.

Turn comparisons occurred at 450 and 350 KIAS, maximum and military power. At 450 KIAS, the F-5 completed a 180-degree turn sooner than the MiG-17, and at 350 KIAS, the two aircraft demonstrated comparable turn rates and radii. In all cases, the F-5 lost considerably more airspeed in the turns than did the FRESCO C. Subsequent offensive and defensive maneuvers illustrated to the pilots of both aircraft that 300 KIAS approximated the airspeed below which the FRESCO C enjoyed a definite turning advantage.

Roll rate comparisons conducted on flight 23 disclosed a wide margin of superiority in favor of the F-5A at 350 KIAS and 450 KIAS. At 350 KIAS, the F-5 rolled 360 degrees and the MiG-17 rolled 180 degrees, and at 450 KIAS, the F-5 completed a 360-degree roll simultaneously with the completion of 90 degrees of roll by the FRESCO C.

Speed brake comparisons were accomplished on both F-5missions, one at constant power, and one at idle power. The constant power deceleration initiated at 400 KIAS and terminated when the MiG-17 indicated 300 knots, at which time the F-5 had 320 KIAS and was approximately 2,000 feet in front of the FRESCO C. At idle power, the advantage of the FRESCO C was less apparent, and over a 130-knot deceleration range, the F-5 was only 500-800 feet in front at termination.

During offensive and defensive maneuvers on three occasions, the F-5 offensively tracked a level breaking FRESCO C long enough to obtain a good gun burst when the FRESCO C airspeed was above 300 KIAS. If the MiG-17 executed an oblique vertical hard turn, however, it rapidly bled airspeed and generated an F-5 overshoot before it adversary could accomplish effective gun tracking.

Defensively, the F-5 could not force an overshoot with a hard/break turn. No further defensive maneuvers were performed; however, as with all aircraft testing on this evaluation, the F-5 could exploit its superior maximum power acceleration potential to prevent a detected threat from arriving at a lethal gun position.

In summary, in a one-on-one situation, the F-5 had to press an attack from a position of advantage until tracking the turning FRESCO C became impossible. At that point, the F-5 had to execute a level or unloaded acceleration. Separation had to ultimately occur, and the F-5 had to retain a high attack speed to facilitate separation. The point at which to execute such a maneuver constituting a difficult decision for the F-5 pilot, since most F-5 pilots were unaccustomed to engaging aircraft with which they cannot turn below 300 KIAS.

Defensively, the F-5 accelerated to defeat a FRESCO C gun attack. If the attack was detected late, a last ditch track spoiling maneuver was required, such as an accelerated, unloaded roll. Such an escape maneuver carried into a steep dive to force the FRESCO C into an undesirable speed realm, if conditions so permitted. It was worthy to note that the F-5 enjoyed an advantage uncommon to the F-100, F-4, and F-105, insofar as it, like the FRESCO, was difficult to see. During separation, it became apparent that the enemy had lost visual contact, at which time re-attack represented a judicious course. The F-5 best positioned for a re-attack with a supersonic, low g zoom to a position of advantage.

The exploitation team did not evaluate flight tactics employed by the F-5 against a FRESCO C threat; however, it seemed logical that the F-5, when numerically superior, defeated the FRESCO C by utilizing split plane maneuvering and mutual support. Such tactics judiciously applied prevented the FRESCO C from achieving a lethal position while the free element brings offensive pressure to bear, thus permitting the F-5 to control the engagement.

The following conclusions pertain to a clean F-5A and clean FRESCO C.

With both aircraft operating at military power, the F-5A possessed an acceleration capability superior to that of the FRESCO C.

With both aircraft operating at maximum power, the F-5A possessed an acceleration capability superior to that of the FRESCO C.

The F-5A zoom capability was superior to that of the FRESCO C.

Below 300 KIAS, the turning capability of the F-5A was inferior to that of the FRESCO. At 350 KIAS, the turning capability of the F-5A was comparable to that of the FRESCO C, and at 450 KIAS, the F-5A possessed a better turning capability had FRESCO. In all cases, however, the F-5A lost considerably more airspeed in the turn, and in the latter two cases, the FRESCO C regained turning superiority or, at worst, reduced the margin of inferiority by employing a, speed brake deceleration in the turn.

The F-5A could not decelerate as readily with speed brakes as could the FRESCO C.

The F-5A rolled at a rate far exceeding that of the FRESCO C and at a vastly superior rate at 450 KIAS and above.

From a position of advantage, the F-5A best defeated the FRESCO C through the employment of hit and run maneuvers.

When defensive, the F-5A defeated an attack by the FRESCO C through acceleration to supersonic speed.

When the F-5 was defensive and the FRESCO C too close to deny it a lethal position by accelerating alone, the pilot accomplished a last ditch maneuver in order to defeat that position. Upon achieving this goal, the F-5A immediately separated, utilizing a steep dive.

The FRESCO C flew at its maximum performance capability throughout the course of this evaluation. The following on was against a single FRESCO in contrast to anticipated multiple FRESCO Cs in combat. Their exploitation results concluded that: flight tactics employed by USAF tactical fighters against FRESCO C type aircraft emphasized mutual support between elements and split plane maneuvering. USAF tactical fighters employ the superior acceleration and vertical maneuvering potential at their command to control engagements with the FRESCO.

Future tactical fighters should consider small or negligible exhaust trails a primary engine design consideration.

Review and update all wing tactical fighter doctrines dealing with specifics concerning FRESCO C engagement criteria to include the findings of this test.

A portion of combat crew training, both initial and recurrency, be devoted to comprehensive instruction concerning aircraft with various performance characteristics.

Continue expending effort in the future acquisition and tactical exploitation of foreign material.

USAF tactical fighters utilize steep dive angle (30 to 90 degrees) escape maneuvers when disengaging the FRESCO C, if circumstances permit.

USAF tactical fighters keep ingress speed sufficiently high to permit rapid acceleration in order to negate a FRESCO C gun attack from the rear hemisphere.

Fighter aircrews remain acutely aware of the visual acquisition problem presented by FRESCO C size targets prior to, during, and subsequent to an engagement.

F-105 vs. MiG-17-F FRESCO C

Fresco C with Normal Russian Markings

Fresco C and F-105

The Air Force and Navy conducted zoom comparisons on flights 7, 14, 28, and 40. With both aircraft utilizing maximum power in the profile described in Test Environment and Procedures, the F-105 averaged a 4,500-foot and 30-knot advantage at termination. The military power zoom was less favorable to the F-105, but still some 1,000 feet in its favor. On that particular maneuver, the F-105 had 20 knots less at termination. During a single zoom

165

comparison with the MiG-17 at maximum power, and the F-105 at military power, the MiG-17F terminated 500 feet above and 200+ knots faster than the F-105.

Turn comparisons performed on flights 5, 6, 12, and 50 demonstrated that at 350 KIAS, the FRESCO C could turn far better than the F-105, and at Mach .9 or 450 KIAS, the results vary, depending on the power settings used. At military power, the FRESCO C enjoyed a slight advantage, and at maximum power, the F-105 turned slightly better; in all cases, however, the MiG-17 retained more energy and terminated at a higher airspeed.

At 350 KIAS, during roll rate comparisons performed on flight 7, the two aircraft rolled at approximately the same rate. At 450 KIAS, aerodynamic forces opposed the deflection of the FRESCO C flight controls, and consequently, the F-105 rolled out of a 360-degree aileron roll before the MiG-17 had reached an inverted position.

Speed brake deceleration comparisons performed on flights 5 and 50 indicated at aver an approximate 100-knot deceleration range, the FRESCO C decelerated 15 to 20 knots more than the F-105.

The exploitation team collected data from the RHAW, radar homing and warning receiver, pertinent to the range only radar of the MiG-17F, and as previously mentioned in the F-4 portion of the discussion. The electronic characteristics of the radar of the FRESCO C illuminated the AAA/AI light of the APR-25/26, and differentiation between that indication and a gun-laying radar indication being difficult in a multi-signal environment. AAA/AI indications underwent careful analyzing before attributing it to an antiaircraft artillery radar.

An evaluation of F-105 offensive and defensive maneuvers employed against a FRESCO C occurred on flights 5-19 12, 14, 28, 29, 40, 41, 49, and 50. Offensively, the F-105 employed yo-yos, lag pursuit attacks, and maneuvers in the vertical plane in attempting to defeat the FRESCO. Hard break turns by the FRESCO C, however, invariably forced the F-105 to overshoot and seek separation.

Defensively, the F-105 could not generate an overshoot through hard/break turns and had to resort to other means to protect itself.

As with the F-4, the exploitation team found that the superior maximum power acceleration potential of the F-105 provided it with the best defensive maneuver against the FRESCO. If, however, detection of the threat did not occur until the MiG-17 had reached a lethal position, some sort of last-ditch maneuver was required for safety during separation.

An unloaded acceleration with subsequent high-calibrated airspeed reversals was the most effective means to that end. Flights 29 and 41 made it apparent that the pilot should keep ingress airspeeds sufficiently high to produce rapid acceleration in the presence of a MiG threat.

In summary, in a one-on-one situation, the F-105 should pursue an attack until tracking becomes impossible, and seek separation immediately, keeping airspeed high throughout the attack. The F-105 cannot afford to engage the FRESCO C in a high angle of attack, slow speed, turning fight.

Defensively, the F-105 employed an afterburner acceleration to nullify an attack by the FRESCO C, and do so in unloaded flight if the threat was from close in. Permitting an escape maneuver to progress into a steep dive angle would benefit the F-105, since it would enhance acceleration as well as force the FRESCO C into high-speed flight. High calibrated airspeed reversals at random intervals effectively produced safety while generating separation. After separation, the F-105 should utilize a very low g turn back to observe the subsequent actions of the FRESCO .If the threat follows; reattack would not normally be judicious.

F-105 flight tactics against the FRESCO C evaluated on flights 12 and 14 utilized split plane maneuvering and mutual support, bringing constant pressure to bear on the FRESCO C. Both elements proved more effective when maintaining high-calibrated airspeeds.

If the F-105 crew could avoid the temptation to track a breaking FRESCO C and separate when an overshoot was inevitable, the flight tactics described could prevent the FRESCO C from ever acquiring a lethal position. Encountering a single FRESCO C required maintaining a constant vigil for another aircraft, and mutual support had to include constant clearing of the rear hemisphere between elements.

The following conclusions were pertinent to an F-105 aircraft configured with AIM-9 pylons on the wing stations and simulated electronic countermeasures pods, and the FRESCO C cleans

• With both aircraft operating at military power, the F-105 had an acceleration capability superior to that of the FRESCO C above 250 KIAS.

• With both aircraft operating at maximum power, the acceleration capability of the F-105 was superior to that of the FRESCO C.

• With both aircraft operating at military power, the zoom capability of the F-105 was slightly superior to that of the FRESCO C.

• With both aircraft operating at maximum power, the F-105 possessed a zoom potential superior to that of the FRESCO C.

• At 350 KIAS or below, the FRESCO C possessed a turning capability superior to that of the F-105 and the margin of superiority increases with a corresponding/decrease in airspeed.

As airspeed approached 450 KIAS, the turning capability of the F-105 approached that of the FRESCO C, but the F-105 lost considerably more airspeed for the same heading change. Further, the FRESCO C could easily regain turning superiority by decelerating in the turn with speed brakes,

When both aircraft employed comparable power settings, the FRESCO C decelerated faster with speed brakes than could the F-105.

The F-105 roll rate with full-scale stick deflection was superior to that of the FRESCO C at high-calibrated airspeeds and comparable at 350 KIAS or below.

From an initially advantageous position, the F-105 should utilize a hit and run maneuver to defeat the FRESCO C.

Defensively, the F-105 could defeat an attack by the FRESCO C by accelerating to supersonic speed. At too close a range to permit immediate safety through acceleration, the F-105 employed a last ditch maneuver to defeat the lethal position of the FRESCO. Seek separation, utilizing a steep dive angle, if possible. The F-105D and F-105F opposed the MiG-17F MiG-17 on flights 5-7, 12-14, 28, 29, 40, 41, and 50. Configurations for the F-105 varied as required to fulfill flight objectives with basic comparison data compiled with the F-105F configured either clean or with AIM-9 rails on the wing stations.

Acceleration comparisons in level and unloaded flight and at maximum and military power conducted on flights 5, 6, 28, 41, and 50 illustrated that at military power, the F-105 possessed an acceleration capability exceeding that of the FRESCO C but not decidedly so until in the vicinity of 275 to 300 KIAS.

At maximum power, the F-105 enjoyed a decisive acceleration superiority over the FRESCO C.

The FRESCO C had a maximum power acceleration capability that slightly exceeded the military power acceleration capability of the F-105.

CHAPTER 7- USN tactical evaluation of the MiG-17F FRESCO C

Fresco C w/USN Attack Airplanes

It is significant to note that the Navy lost every simulated air combat exercise on their first sortie against the HAVE DOUGHNUT MiG-21 FISHBED, HAVE DRILL, and HAVE FERRY MiG-17F FRESCOs.

Navy A-4, A-6, A-7 air attack aircraft failed to gain the offensive on the FRESCO C in any flight regime throughout the tactical phase of the named exploitation projects. At CRT, the FRESCO C out accelerated, out climbed, and out zoomed all Navy attack airplanes. Below 475 KIAS, the FRESCO C enjoyed a marked advantage in sustained turn performance. Maximum steady state roll rate was the only area of advantage for the attack planes. The FRESCO C excelled in the slow speed regime. The A-7 made two reversals in a scissors maneuver prior to the FRESCO C reaching a gun tracking position. The A-4 and A-6 completed only one reversal prior to losing the slow speed fight. None of the attack airplanes gained an offensive position on the FRESCO C.

The Navy attack airplanes A-4, A-6, A-7 air attack aircraft successfully disengaged from the FRESCO The superior roll rate and better high j maneuverability of all Navy attack airplanes permitted them to disengage and escape from the FRESCO C. With the FRESCO C in a gun tracking position, unloaded maximum roll rate reversals, while accelerating to maximum q, negated the FRESCO's tracking solution immediately and placed the attack airplanes outside maximum gun range. Of particular advantage to the attack airplane pilots was their familiarity with high-speed low altitude maneuvering. Because of the flight control problems, the FRESCO C pilots were extremely uncomfortable attempting to follow attack airplanes down to tree top level at speeds in excess of 475 KIAS. The FRESCO C pilots usually refused to follow the attack airplanes into this regime. The A-6, with a higher q limit, escaped with more facility than either the A-4 or A. Once the FRESCO C achieved a gun tracking position, a breaking turn by the attack airplanes failed to cause an overshoot.

Losing 100% of their first encounters against the MiG in Nevada turned the Navy's focus to human factors and personnel evaluations. It is important to note that overconfidence was an important derogating factor in the performance of US Navy pilots while fighting the FRESCO C for the first time. Prior to fighting the FRESCO, there was a tendency to underestimate its performance. Although last produced in 1961, the FRESCO was still a formidable threat to all Navy tactical airplanes.

Aircrews remained constantly vigilant and were not lulled into complacency by the age of the FRESCO. Pilot performance improved dramatically after the first engagement with the FRESCO. All project pilots agreed that prior to their first engagement; they had seriously underestimated the capabilities of the FRESCO. Though subsonic, the FRESCO was a dangerous adversary. Substantial information concerning the actual capabilities of the FRESCO C was available prior to Project HAVE DRILL. In all significant cases, the results of the project validated this information. It was felt that underestimating the FRESCO was due to its vintage 1950 and simplicity as compared to its opponent, i.e. F-4, F-8, etc, and not the published performance characteristics.

Due to system simplicity and the absence of dangerous flight characteristics, the pilot employed the FRESCO to its fullest extent, finding the airplane easy to fly. The pilot could not exceed the structural limit load factor of 11 g. The lack of abnormal stall or spin characteristics allowed the aggressive pilot to approach the edges of the FRESCO's flight envelope with confidence. The weapons system simplicity eliminated a good deal of pilot confusion presented when utilizing systems that are more complex. After minimum training, the pilot realized that he could not over stress the airplane, could quickly recover from uncontrolled flight, and effectively employ the weapons system. This overall simplicity and ruggedness were of great significance in the ACM environment.

Like the Air Force, Navy attempted with its flights to duplicate the ACM environment encountered in SEA. However, unlike true combat, the Navy choreographed its exploitation program with the other participants to include briefing the participating pilots prior to each flight, maintaining two-way UHF radio communication among participating pilots, and terminating engagements when encountering unusual flight characteristics.

The participants strived to assume a covering Fighter/Engaged Fighter posture in every engagement. If attacked from aft of abeam while in Combat Spread, they unloaded for airspeed and separation (500 KIAS minimum). They increased vertical and lateral separation between their airplanes by the outside man going up and outboard, applying g only to establish a climbing attitude. If the MiG-17s split, the Navy pilots continued to accelerate and extended the fight beyond their q limit. When approaching 600 KIAS, the inside pilot started a slight turn away from his wingman who then maneuvered for an attack on the inside MiG-17. The separation between the outside friendly and the outside MiG combined with the high airspeed prevented his re-entering the fight before his wingman came under attack. If the FRESCOs choose to stay together on one friendly, the Navy pilots dragged them out to their q limit while increasing vertical and lateral separation between fighters. At this point, the engaged fighter (the one the MiGs chose to chase) performed a flight turn away from his wingman who turned in for his attack as the MiG started to follow him around. The covering fighter avoided bleeding off speed to below 475 KIAS and maneuvered into the MiG-17's blind area, i.e., below the plane of the MiG-17's wings. If the FRESCOs decided to slow down and turn, the covering fighter unloaded and separated, attempting to drag the MiGs after him. At this point, he called his wingman in to assume the covering fighter role.

If attacked by a MiG-17 one-on-one, they did not try to force an overshoot. Instead, they sacrificed angle off to gain nose to tail separation, while accelerating beyond the FRESCO's q limit, jinking (changing the flight path of the aircraft in all planes at random intervals) while accelerating to destroy his gun tracking solution. During separation, they maintained visual contact and avoided arcing excessively to prevent the FRESCO C cutting across their circle and getting a shot. As the range approached visual detection limits (2-3 nm), they performed a maximum rate, minimum radius reversal to meet the FRESCO head-on while striving for maximum efficiency of their plane. Knowing the FRESCO's performance being contingent upon use of afterburner, the Navy stressed winning the fight by having fuel left when the MiG pilot had to bingo.

The Navy's strategy exploited the FRESCO's slow roll rate and blind cone below the plane of his wings. This required the FRESCO pilot to counter with a roll where the MiG-17's poor roll rate worked against it. Knowing the MiG-17's performance declared severely above 475 KIAS, the Navy fighter planes attempted to fight the MiG-17 at energy levels exceeding the capabilities of the MiG-17.

The A/B on the FRESCO engine gave it a performance level impossible to duplicate or realistically simulate by US airplanes with similar turn capability. Thus, even with 2 1/2-3nm separation, it required a great deal of technique to turn for re-engagement. If an aggressive FRESCO pilot saw an attack by Navy fighters in the aft hemisphere, he could force an overshoot at all airspeeds below 475 KIAS. The Navy fighters learned to not slow down and turn with the FRESCO, executing a high Yo-Yo as soon as practicable. A roll away during the Yo-Yo was an effective lag pursuit maneuver and also put the attacking fighter in an area difficult to see (i.e., high and aft) and one where the FRESCO pilot did not expect the fighter to be. This method of countering the overshoot normally increased the longitudinal separation between the FRESCO and fighter and, thus, aided in weapons system employment. The reversal at the top was normally a rolling split-S (bank angle about 135 degrees). Most importantly, this method helped preclude trapping the fighter into a very close-in turning duel.

Classic loose Deuce maneuvering was a great aid in maintaining mutual support with the greater separations required to effectively combat a section of FRESCOs. Aircrews pressing for the "quick kill" made mutual support difficult and also run the risk of allowing the fight to degenerate into a turning duel.

Due to the great performance advantage enjoyed by the F-8 and F-4 over the FRESCO, the used of full A/B was not required as frequently or for such great intervals of time, as when fighting a FISHBED or other similarly performing airplanes. Fuel was one of the FRESCO's prime constraints and severely limited its offensive capability.

The tactical engagements over Groom Lake established that A-4, A-6 and A-7 airplanes should not attempt to engage the MiG-17 in ACM, and if engaged, escape expeditiously by utilizing superior roll rate, unloaded accelerations, and high-speed, low-level run outs. When a sighting a FRESCO making an attack on you, go to full power and immediately jettison all external stores while creating as much angular separation as possible while staying unloaded and diving for airspeed. If the FRESCO closed to a gun firing position, perform a maximum rate roll away from him, staying unloaded and continue your acceleration. Hold this position for 1-2 seconds and then roll at maximum rate back toward the FRESCO. If you kept g off the airplane and continued your unloaded diving acceleration, the range opened significantly. Take the FRESCO down to as low a level above the ground as possible, maintaining your airspeed as high as possible. The FRESCO pilot would be unable to attack you, low level, at speeds above 500 KIAS. If necessary to engage the FRESCO, utilize the fighter offensive separation tactics, described earlier in this section to exhaust the FRESCO's fuel supply. Take extreme care under these circumstances to avoid other enemy aircraft that could ch00se to enter the fight.

It was the opinion of the MiG-17 pilots that the vulnerability to combat damage of the FRESCO was quite low. The FRESCO was a rugged and durable airplane. Since the VK-1F was a centrifugal flow engine, its vulnerability to gunfire or fragment damage was much lower than that of the axial flow engine. The MiG-17 only hydraulically boosted the ailerons and provided manual backups with mechanical pushrods provided for all controls, which also lowered the kill probability from gunfire or fragmentation. All fuel cells were located in the fuselage. Thus, the wings were relatively invulnerable. Armor plating protected the pilot.

The Navy concluded the FRESCO C capable of defeating any USN tactical airplane in a turning fight at speeds of 475 KIAS and below. In the subsonic region, the FRESCO C, properly employed and aggressively maneuvered, was a highly effective air superiority weapon system.

The FRESCO C was simple and reliable enough to operate from remote sites with a minimum of support equipment. The subsonic maneuvering capability and overall performance of the MiG-17 had been grossly underestimated. One could fly the MiG-17 with confidence to the limits of its operating envelope. The gun system of FRESCO C was highly reliable and effective The FRESCO C had lower vulnerability to combat damage than most current US Airplanes. The FRESCO C had good visibility above its horizontal plane but poor visibility below its horizontal Plane. Camouflage paint and the small size of the airplane made visual acquisition and retention of the FRESCO C difficult in the air-to-air environment. Due to fuel limitations, the FRESCO C cannot engage in pro-longed ACM beyond 75 m of its home base. (FTD felt statement had to be qualified.) The q limit of the MiG-17 seriously inhibits its total effectiveness in ACM. The FRESCO C was extremely fuel limited. All USN tactical airplanes had higher q limits than the MiG-17. All USN tactical airplanes had higher roll rates than the MiG-17. The head-on VID as currently flown in the 1-4 was not effective against the FRESCO because it was dependent on full system radar track by both attacking airplanes and sacrifices tactically advantageous positioning. USN attack airplanes had no offensive capability against a properly flown MiG-17. The oblique loop maneuver used against the MiG-21 was ineffective and dangerous against the MiG-17. Due to overall performance superiority, two properly flown F-4's could remain 100 percent offensive against two MiG-17's in air combat, Padlock technique was mandatory during ACM with the MiG-17.

MIG-17 with F-4 and F-8

The MiG-17F HAVE DRILL & HAVE FERRY WITH NAVY F-8

CHAPTER 8 - MiG-17F FRESCO exploitation conclusions

Navy recommendations

Commander Operational Test and Evaluation Force recommended that: Navy fighter airplanes utilize the following tactics to defeat the FRESCO C.
- Maintain a high energy level-500-600 KIAS.
- Avoid high g maneuvers below 500 KIAS.
- Use thrust advantage to prevent the FRESCO from attaining a lethal gun tracking position.
- Force the FRESCO to fight at airspeeds above 475 KIAS.
- Engage only as a section with strict mutual support, utilizing "loose Deuce" maneuvers and striving to maintain an offensive fighter posture.
- Use A/B judiciously and efficiently to take maximum advantage of the FRESCO's limited fuel supply.
- Exploit the FRESCO's weaknesses i.e., blind area below the horizontal plane, poor roll rate and marginal control ability in the high q regime.
- Utilize when necessary, maximum rate, minimum radius turns at ranges to the FRESCO of 2-3nm, and base the decision to reverse on respective energy levels and the tactical situation.
- A-4, A-6 and A-7 airplanes not engage in ACM with the FRESCO C.
- If engaging the FRESCO C with A-4, A-6, and A-7 airplanes, disengage by jettisoning all stores not possessing air-to-air capabilities, performing an unloaded acceleration while diving to low level, by utilizing maximum roll rate, and running out at maximum airspeed and minimum altitude.
- If a reversal or turn was necessary, insure that the range to the FRESCO was greater than 2-3nm, then performed a minimum radius and maximum rate turn to passed the FRESCO head-on.
- Maintain strict lookout at all times in hostile territory.
- In a threat area, weave and vary headings along the basic course•
- Practice realistic AC25 as much as possible, under controlled conditions, against small airplanes with low wing loading.
- Whenever possible, conduct radar intercept and training over land.
- Pilots practice coaching RIO's on to targets at all times, even when not engaged in ACM.
- Intensify squadron level training of RIOs in the area of air combat tactics.
- Intensify ACM training of attack aircrews at all levels.
- Pursue continued exploitation of foreign airplanes.

Tactical flight summaries

Summarizing the HAVE DRILL and HAVE FERRY exploitation efforts, the exploitation found the MiG-17F easy to fly, and the undesirable features, such as spin and accelerated stall, readily handled when encountered. With proper user knowledge of its maneuvering capabilities and limitations, the exploitation team found the MiG-17F a very effective interceptor/air superiority daylight fighter throughout most of the subsonic flight envelope.

The MiG-17F demonstrated outstanding lift-limited maneuvering capabilities. Available g and g-level for onset buffet were high and turn radius low. Thrust limited turning performance was also good. Longitudinal stick forces increased significantly beyond .85 Mach number and were excessive above .90 Mach. Lateral-directional damping was generally weak and especially poor near .92 Mach Number Lateral-directional tracking for gunfire suffered from turbulent weather or small, inadvertent control disturbances,

Roll rate capability was low, only 100-130 degrees per second. Sufficient aileron deflection was not available past 572 KIAS to prevent the airplane rolling off.

The exploitation team found the MiG-17F very reliable and its operational availability exceptional. The exploitation team easily accomplished four to five flights day after day with briefings, debriefings, and turn around times of other participating aircraft being the only factors limiting additional flights. Considering these conditions, the reliability record of the MiG-17F, during the evaluation, was not only exceptional but was a sobering fact.

The MiG-17F was a very easy aircraft to maintain. Soviet design approach during the manufacture period of this aircraft was toward a rugged airframe, engine and system simplicity and overall reliability. Attractive features included dependable in-flight gun clearing and charging, and a smooth, well-balanced flight control system. Particularly impressive was the lack of engine exhaust smoke at lower altitudes.

The MiG-17F's simplicity and rugged design lent it the capabilities to operate from unprepared landing strips and for full maintaining by underdeveloped nations, thus allowing it to be a continued threat during limited war situations.

The MiG-17F design concentrated the vulnerable critical flight components in a small volume thus presenting a small vulnerable target. The survivability design philosophy of this aircraft protected these vulnerable components by masking and armor from front and rear. Vulnerability to continuous rod warheads, typical armaments used in the AIM-7E, AIM-90, and AIM-54A, was slightly lower than previously assumed by the ordinance community. Factors pointing to this assessment liberal used of high strength steel components throughout the aircraft and the push-pull rod system used for control actuation.

The VK-1F engine of the MiG-17F performed close to factory specification during the HAVE DRILL program. Exhaust gas temperature, thrust and fuel flow were all within the manufacturers prescribed operating limits. The maximum physical RPM at which the exploitation team ran the engine was on an average 1.12 percent below specification value. Engine pressure ratio at military power was in exact agreement with the Soviet published data.

The reader should note that the author bases the performance data presented in this book upon the individual aircraft studied. The author attributes differences between test data and published estimates to non-standard day conditions, uncorrected engine performance, and/or incorrect comparisons between flight rules or profiles. In most cases, comparisons were representative of typical aircraft flight-testing results utilizing all sources of information for correctly and analyzing completely the weapon system capabilities.

Both programs wound down by June 1969, with the findings shared with the Navy's new TOPGUN School established after these exploitation programs revealed the need to reintroduce dogfighting skills to Navy pilots. The Navy shared the findings of HAVE DOUGHNUT, HAVE DRILL, and HAVE FERRY with the instructors at the USAF's Fighter Weapons School at Nellis

AFB, Nevada to later induce the Air Force to establish the Red Flag aerial combat training exercises mimicking real air combat situations.

Continue USAF efforts to acquire threat type fighter aircraft for exploitation. This project provided urgently required information for AD. This command presently had a unit deployed to Korea where the vast majority of the threat was from the MiG-17. As a result of this project, that unit and future deploying units were better prepared to complete their flights. Information gained from "HAVE DRILL, HAVE FERRY" and future foreign aircraft exploitation insured development of adequate tactics to counter those aircraft and design of future US fighters with superior performance, systems, and armament. Aerospace Defense Command remained prepared to participate in the exploitation of foreign aircraft to insure the combat effectiveness of US military forces.

Summary of delayed discrepancies and required maintenance generated as a result of the project.[11]
- VHP radio removed
- IFF not installed
- Glare shield bracket removed
- VHF power supply removed
- All ammunition cans removed
- VHF control head removed
- Radio compassed indicator removed
- Left wing IFF crystal removed
- Ballast installed in nose bay (84 lbs)
- Right wing IFF crystals (2 ea,) removed
- Tactical bombing system inoperative, 10 AMP fuse removed
- SIRSITA bell in cockpit removed
- Afterburner flameholder distorted and cracked
- APT section removed to facilitate other maintenance
- Afterburner assembly removed
- Two holes melted through heat shield in APT section
- Bottom afterburner eyelid actuator APT mount chaffing AFT section
- Hydraulic line to bottom afterburner eyelid actuator chaffing AFT section
- Bottom rib in APT section below bottom afterburner actuator cracked and distorted
- Brakes weak

Maintainability (HAVE DRILL)

The following figures presented the manhours expended for maintenance on the HAVE DRILL MiG-17 along with some sortie information.
- 498.9 Manhours for discrepancies
- 89.9 Manhours for periodic inspections
- 86.0 Manhours for postflights and preflights
- 674.8 Total manhours for maintenance
- 131.3 Total flying hours.172 Total sorties

---

[11] The exploitation team discovered the last eight discrepancies when the project terminated.

- 5.1 Manhours per flying hour 3.9 Manhours per sortie
- 101 Maintenance discrepancies of these, 30 were on radar
- 498.9 Manhours for discrepancies of these, 259-7 M.H. were on radar
- 55 Plying days.3.1 Sorties per flying day

The aircraft turned around several times in 30 minutes. This included 10 minutes towing time to the end of runway prior to take-off.

MiG-17 Deficiencies noted

Above Mach .85 or 450 KIAS, whichever was lower, the FRESCO C encountered aerodynamic forces that oppose deflection of the flight controls, resulting in a very slow available roll rate and pitch change in that speed realm.

The FRESCO C flight controls lacked dampening in any axis. Above 375 KIAS, the aircraft possessed a Dutch Roll tendency. At any speed in turbulent air, the aircraft was difficult to control in yaw.

The following deficiencies limited the FRESCO C.
- Low firing rates of the 2 mm, and 37 mm cannons (900 and 400 rounds per minute, respectively).
- Low muzzle velocity of the 23 mm and 37 mm ammunition (2,250 fps for both calibers).
- Excessive tracking time for proper lead computation (two to three seconds).
- Total lead required due to low muzzle velocities and projectile weights.

The exploitation team concluded from these flights that flight tactics employed against the FRESCO C by all tactical fighter aircraft that participated in this evaluation should emphasize mutual support between elements and split plane maneuvering.

Prolonged high angle of attack maximum performance maneuvering and low airspeed (250 KIAS or slower) defined a performance envelope wherein the FRESCO C was superior to all tactical fighters that it confronted on this test.

The F-4, F-105, F-100, and F-5 aircraft all possessed a vertical maneuvering superiority over the FRESCO that the case of the F-4 and F-5 aircraft provided a decisive advantage with low g pitch movements to the vertical plane.

When the FRESCO C engaged any of the aforementioned tactical aircraft at calibrated airspeeds in excess of 450 KIAS, USAF aircraft defeated its lethal position by utilizing rapid unloaded reversals and check turns of 30 to 60 degrees while maintaining high-calibrated airspeeds.

USAF tactical fighters could, by accelerating in nose low attitudes of 30 to 90 degrees during an escape maneuver, force the FRESCO C pilot into a realm of flight wherein his capability to pull out becomes his dominant consideration.

The F-4 was by far the easiest of the participating aircraft to acquire visually, followed by the F-105 and F-100, in that order. Both the FRESCO C and F-5 proved to be very difficult to acquire with the unaided eye. Most first encounters with the FRESCO C demonstrated that highly qualified pilots made serious misjudgments in range estimation.

Incorporate the findings of this test into the tactical doctrines of Southeast Asia fighter wings for more accurate instructions for FRESCO C engagements.

The APH-25/26 of the F-4 and F-105 furnished an indication of the presence of the FRESCO C Scan Fix, range-only radar, accompanying the strobe by the illumination of the

AAA/A1 light.

The MiG FRESCO C aircraft was relatively easy to fly and required minimum time to obtain an average proficiency level.

At medium and low altitudes, the aircraft appeared to operate best at speeds approximating 300 to 350 KIAS.

Its acceleration potential of low speeds (180 KIAS or less) to 250 KIAS provided a significant advantage to the FRESCO C in a low-speed maneuvering engagement.

At airspeeds of 350 KIAS or below, the FRESCO C possessed an outstanding turning performance.

Above Mach .85 or 450 KIAS, whichever was lower, the FRESCO C encountered aerodynamic forces that opposed the deflection of the flight controls; hence, in that speed realm, roll and pitch rates deteriorated markedly.

The aircraft was extremely small and difficult to acquire visually, a characteristic enhanced by a negligible exhaust trail.

Maximum speed of the aircraft at low and medium altitude in level flight was approximately Mach .96, but high stick forces prevented maximum performance maneuvering in that speed range.

The bubble canopy, with a mounted periscope assembly, afforded the FRESCO C excellent visual acquisition in the rear hemisphere. Poor visibility looking low, both forward and laterally, and the periscope assembly interfered somewhat with overhead visibility.

When configured clean with full internal fuel, the FRESCO C had approximately 373 gallons °F fuel available to the engine. Maximum performance maneuvering at medium or low altitude consumed the available fuel available in 20 to 25 minutes.

The FRESCO C flight controls lacked dampening in any axis. In turbulence, control of the aircraft was difficult in the yaw axis. At 375 KIAS or above, it demonstrated Dutch Roll tendencies or instability in all axes.

When the FRESCO C employed against aerial or around targets, the weapons system was effective under ideal conditions. However, low cannon firing rates, low projectile muzzle velocities, excessive lead computation time, excessive requirement for lead, and excessive yawing motion in turbulent air limited the overall effectiveness of the weapons system.

TAC summary of the test program of the MiG-17F FRESCO C

Prior to commencing the tactical phase of exploitation, the two primary TAC pilots flying the majority of the total TAC flights (52 of 57) checked out in the FRESCO C in two flights and one flight, respectively, prior to the initiation of the TAC test program. A third TAC pilot received a two-mission checkout program on TAC flights 38 and 39. All three TAC pilots expressed the following opinions regarding the operational characteristics of the MiG-17:

They encountered no difficulty in starting the aircraft.

The aircraft was somewhat cumbersome to taxi. This fact arose, in part, from the peculiar mechanics involved in braking, and utilization of differential braking for directional control. Applying hand pressure to a level on the control stick column controlled air brake pressure, and deflecting the rudder pedals accomplished differential braking.

Once airborne, the aircraft exhibited characteristics that rendered the pilot comfortable at the controls. All three pilots felt capable of operating the aircraft at maximum performance after concluding their initial checkouts.

Throughout the majority of the flights on which the aircraft was flown, 300 to 350 KIAS appeared to be the optimum airspeed range for maneuverability and weapons employment (reference flights 1-9, 26, 27, 34, 35, 36, and 37). Below 300 KIAS, the turning rate and radius of the aircraft improved, as did its acceleration performance, but a loss of sustained maximum performance maneuvering capability impart offset these advantages. Above 350 KIAS, other considerations arose that would be amplified upon as this discussion progresses.

Straight, level, and unloaded acceleration comparisons at both military and maximum power favored the USAF tactical fighters at high-calibrated airspeeds. The FRESCO C had an excellent acceleration potential near 250 KIAS, however, and frequent mention occurred in the summaries of no significant advantage befalling USAF fighters until airspeeds approached or, in the case of the F-100, exceeded 300 KIAS. On TAC flight 1, initiating a military power acceleration comparison with an F-4 at 250 KIAS enabled the FRESCO C to exceed the acceleration of the F-4E prior to 260 KIAS. During the evaluation of certain offensive and defensive maneuvers, the FRESCO C at low calibrated airspeeds displayed a consistent ability to out-accelerate the participating fighters.

The exploitation team conducted turn performance comparisons on numerous flights in the course of this evaluation at speeds of 350 KIAS, 450 KIAS, and Mach .9. At 350 KIAS, the FRESCO C enjoyed a turn superiority over the F-100, F-105, and F-4, and realized a parity with the F-Decreasing airspeeds from that point resulted in an increase of the margin of superiority against the three aircraft mentioned, and at 300 KIAS the MiG-17F could out-turn the F-5. Increasing the airspeeds from 350 KIAS, however, presented a somewhat different situation. At about Mach .85 or 450 KIAS, whichever was less, the unboosted rudder and elevator of the FRESCO C encountered aerodynamic forces, which opposed the deflection of the control surfaces. The exploitation team noted this tendency on the initial TAC flight, and as a characteristic of the aircraft throughout the evaluation, consequently, referring to the turn comparisons conducted at 450 KIAS and Mach .9 not altogether impressive. Another factor then presented itself, however, and that was the airspeed lost in the execution of a maximum performance turn. Although at high-calibrated airspeeds, the F-105, F-100, and F-5 completed a

180-degree turn prior to the time that the FRESCO C could do so, and the F-4 matched turn rates with the FRESCO, the MiG-17 invariably lost less airspeed in the turn.

A canopy mounted periscope assembly enhanced visibility from the cockpit in the rear hemisphere over the pilot's head. This assembly afforded visual lookout of approximately 20 degrees to either side ©f the tail, and 10 degrees up or down. Overall visibility was not good. The seat sat low in the cockpit, and even when topped, the pilot experienced difficulty seeing over the nose and sides. Additionally, the periscope assembly itself obstructed overhead visibility and inconvenienced the pilot. The FRESCO C was extremely small and very difficult to see, particularly if one was unaware of its presence. It did not emit a smoke trail of any significance whatever, which complicates the visual detection problem even more. It was interesting to note the short (1-3 HM) visual acquisition ranges and positive ID ranges recorded by 1 &-4 crews on flights 33 34, and 35, and to contrast those ranges with the ranges at which the FRESCO C pilot was able to visually acquire the F-4s (10-15 nautical miles).

The FRESCO C had undampened flight controls. The aircraft became unstable in all axes at 375 KIAS and higher and directional control was poor thereafter. This characteristic, encountered in TAC flight 1 and noted throughout the test became significant when the aircraft attempted to track an airborne or ground target at that speed. Further, yaw control of the aircraft became difficult when the airflow over the flight controls was disturbed. Consequently, the aircraft was unstable in yaw at any speed in turbulent air

The FRESCO C flew in a clean configuration on all of the 57 TAC flights flown. This configuration yielded flight durations ranging from 25 to 45 minutes, depending on the type flight flown. Obviously, a flight requiring frequent resort to utilization of maximum power enabled less time aloft than did a flight requiring lesser power settings. Sustained low altitude maximum performance maneuvering by the FRESCO C exhausted the available internal fuel in 20-25 minutes. It was, therefore; apparently better to utilize the aircraft in a point defense role, wherein the fuel consideration was not prohibitive.

Maximum speed of the FRESCO C at maximum power in level flight, according to FTD-CS-O9-5-67, was about Mach .96 in level flight. For effective employment as a weapons system, the FRESCO C should not fly at high-calibrated airspeeds because of the previously mentioned high stick forces affecting the maneuverability of the aircraft.

The speed brakes of the FRESCO C were very efficient. Evaluated in comparison to those of the participating aircraft on flights 1, 8, 22, and 50, they varied by comparison; but in all cases, were more effective than those of the other participants were. This fact remained true, regardless of the power settings utilized with comparable power settings maintained throughout in both aircraft.

The FRESCO C flew and evaluated against a towed TDU-10B dart on six different occasions in addition to four flights dedicated to the evaluation of the FRESCO C in an air-to-ground strafing role. The exploitation team obtained good results under ideal conditions (low attack speed, smooth air, low g), however, at less than ideal conditions (higher attack speed, turbulent air, higher g), the results were less favorable. Numerous considerations detracted from the effectiveness of the armament. As previously discussed, the exploitation team found the FRESCO C unstable in turbulent air and at airspeeds greater than 375 KIAS.

The exploitation found the MiG-17 to have low firing rates. The two 23 mm cannons fired at the rate of 900 rounds per minute, and the 37 mm cannon fired at a 400 rounds per minute rate. With all three guns simultaneously fired, thus the total cumulative firing rate equated to 2,200 rounds per minute, a low rate with compared to the 6,000 rounds/minute rate of the Gatling gun on the F-4 and F-105.

The exploitation found the MiG-17 to also have a low muzzle velocity. The muzzle velocity of both projectile sizes was 2,250 feet per second. By comparison, the muzzle velocity of USAF 20 mm projectiles was 3.280 feet per second.

To meet the excessive tracking requirement, the optical sight provided a proper lead computation after two to three seconds of steady state tracking. It was possible that the dynamics of an aerial engagement could yield that requirement prohibitive.

The low muzzle velocities of the projectiles and their weights precipitate a requirement for a large amount of lead as compared to the USAF 20 mm weapon. In a maximum performance maneuvering engagement, the extent of lead required could exceed the ability of the aircraft to obtain it.

Two sources of FRESCO C performance data were available to assist the project officer in the conduct of this test, and they found to be slightly divergent. The sources were FTD-CS-O9-5-67, FRESCO (MiG-17) Weapon System, 18 July 1967 (S), and APGC-TR-66-4, dated March 1966 (S). The former more accurately described the performance characteristics of the test bed aircraft after the AFFTC completed a comprehensive flight test evaluation of the aircraft.

The aircraft flown on this evaluation proved to be highly reliable and maintainable. Twenty-three major discrepancies occurred on 224 flights between 1 February 1969, and 14 March 1969 and in that period with only two flights lost to maintenance. These figures were even more impressive considering the fact that the aircraft frequently flew four flights in a single day.

CHAPTER 9 - Other MiG Projects in Nevada

Air Force assuming CIA operations at Area 51

SECRET

DEPUTY DIRECTOR OF CENTRAL INTELLIGENCE
WASHINGTON, D.C. 20505

26 August 1976

C/S has noted
27 AUG 76

General David C. Jones
Chief of Staff
United States Air Force

Dear General Jones:

You will recall that in an exchange of messages between you and Mr. Colby in February and March 1975 (OMAR 8020 and PADRE 5406) you expressed the desire to have CIA continue to manage Area 51 at the Nevada Test Site. Mr. Colby agreed to retain management of the Area "as long as community intelligence requirements and budgetary constraints permit." Subsequently, the Senate and House Appropriations Committees deleted Area 51 operating funds from the FY-76 CIA appropriation, but in accordance with USAF desires, recommended that CIA continue to manage the Area with the operating funds being provided by the Air Force. That recommendation was implemented, and we have been operating under the arrangement for a little over a year.

In a recently completed review of the operation of the Area, including the projects currently underway there and those planned for the future, it was concluded that management of the Area by this Agency is no longer warranted. Essentially all of the activity at the Area is in support of DOD projects, the Air Force's HAVE GLIB project being by far the largest at the present time. Another large Air Force project and a large Army project are projected for the near future. CIA's use of the Area for its own projects is minimal, and we see no significant increase in our requirements for the future. Although it is true that the DOD projects are intelligence related and do require a covert base, I am convinced that the Air Force could manage the Area and maintain its covert posture. In addition, it would appear to be better management practice to have the funding organization responsible for the operational management of the Base.

This issue was brought up before the Committee on Foreign Intelligence (CFI) as a key topic in the review of the Agency's FY-1978 program. The CFI approved the recommendation that management of Area 51 be transferred from CIA to the Air Force by FY-78. We therefore plan to begin a gradual phase-out that will be complete by the end of FY-77. We hope that by beginning now we can effect a smooth transition by that time with little or no effect on tenant projects.

Released in October 2013

Sincerely,

E. H. Knoche

18 72:76

E-2 IMPDET
CL BY 034944

Project HAVE PRIVILEGE

On 25 November 1969, a Cambodian Khmer Air Force pilot defected to South Vietnam in the Chinese copy of the MiG-17F, the Shenyang J-5. The AFSC/FTD tapped Col. Wendell Shawler, the USAF pilot, who flew the MiG-17F in HAVE FERRY and HAVE DRILL, to go to South Vietnam and made several evaluation flights of the J-5 to establish it having the same flight characteristics as the MiG-17. This short program of just five flights from Phu Cat AB in South Vietnam was code named HAVE PRIVILEGE.

Resulting from these four top-secret exploitation programs, both USAF and Navy fighter tactics were changed and pilots were once again trained to exact as much capability and performance out of the aircraft as possible to win the dogfight. In 1989, a Pentagon official finally confirmed the 1981 combat of two US Navy F-14 Tomcats versus two Libyan Sukhoi Su-17 fighters over the Gulf of Sidra used tactics developed out of mock combat testing with US-operating Soviet fighters.

At least one more MiG-21 captured in Vietnam and a MiG-25 flown by defector Soviet Pilot Viknotsor Belenko arrived at Area 51 where they flew before returned to the source. In 1973, the exploitation continued with Project HAVE IDEA, which took over from the older HAVE DOUGHNUT, HAVE FERRY, and HAVE DRILL projects. In July 1975, the 4477th TEF ("Red Eagles") formed at Nellis AFB and in December 1977, the 6513th Test Squadron ("Red Hats") formed at Edwards AFB to continue testing these foreign craft. Some aggressor training was done where the units went head to head against USAF fighters in mock dogfights now to find out and exploit possible weaknesses.

Project HAVE PAD

In 1978, the AFFTC, TAC, and the US Navy exploited the Soviet MiG-23 Flogger at Area 51 in a project codenamed HAVE PAD. The primary objective of HAVE PAD was to determine the performance and flying qualities of the MiG-23MS with a secondary objective of conducting a tactical evaluation to determine its combat effectiveness against US aircraft.

The FLOGGER MiG-23MS, powered by a R29-300 power plant, was an export fighter-interceptor capable of a dual role in air-to-air intercept or ground attack. It was a versatile aircraft because of its inherent interceptor design, variable geometry wings, AI radar, multiple weapon stations, navigation and communications equipment (RSBN and data link), high top speed (Mach 2.35, low landing speed (140 knots), and excellent acceleration characteristics. In addition, it had a maximum endurance capability of 4.3 hours cruising at 10 km altitude at Mach .65 using 3/800 liter external fuel tanks. Its maximum operational range was 1080 nautical miles without external fuel and 1565 NM with 3 x 800 liter fuel tanks.

Access to the MiG-23 weapon system made its exploitation important to the US Air and Navy who believed the MiG-23 an excellent dogfighter. Evaluation of the AI radar was also important as the Soviets now used it in the late model MiG-21 FISHBED planes, making them a quantitative and qualitative threat. The subsystems of the 1974 production MiG-23 represented the heart of nearly every Soviet late model domestic and exporter fighter aircraft flying at the time. The technology benchmark recorded in this plane coupled with the earlier MiG-25 exploitation provided a unique database upon which to better judge current and future Soviet capabilities. Its rugged tricycle landing gear allowed operation on both concrete and sod airfields.

The exploitation team found that despite its small size, excellent engine response, nearly jam-proof fire control radar system, and gun/missile weapons load, the FLOGGER was not an extremely capable dogfighter. They found its turning performance unimpressive, cockpit visibility was poor, the engine smoked, flight control forces were high, and the variable wing sweep system was neither optimized nor automated. Pilot aids were "head-down" and not optimized for the aerial combat arena.

The exploitation of the HAVE PAD plane's avionics systems afforded an unparalleled opportunity to examine, test, and photograph current Soviet avionics equipment over a long period. This yielded some very significant results. The ARK-15M radio compass was the first military avionics system to have integrated microcircuits in two-sided, printed circuit board, plug-in cards, and digital processing. Testing the JAY BIRD AI radar confirmed the operating frequency range of 12,730 - 13,220 MHz and the search/track ranges of 30/20 km. The JAY BIRD was very similar to the I-band portion of the FOX FIRE in configuration, physical appearance, and operation. Peak power was about 170 kW, which was higher than the 100 kW previously estimated. The JAY BIRD contained an impressive collection of ECCM features, making it highly resistant to conventional jamming techniques. The testing of the rate gyros of the SAU-23 automatic control system in the laboratory at Groom Lake made it the first exploitation of Soviet aircraft gyros.

The armament complement of the FLOGGER was a good mix for a fighter-interceptor with the internal gun, ATOLL missiles, and a variety of air-to-surface weapons. Unfortunately, no ATOLL missiles came with the plane, denying the exploitation team on opportunity to exploit the semi-active (SA) missile, leaving a gap in knowledge of a significant threat item. However, the exploitation team found weak points in the FLOGGER weapons area that included the inability to launch single bombs or rockets, poor switching requirements, no head-up display, and no off-axis missile lock-on capability.

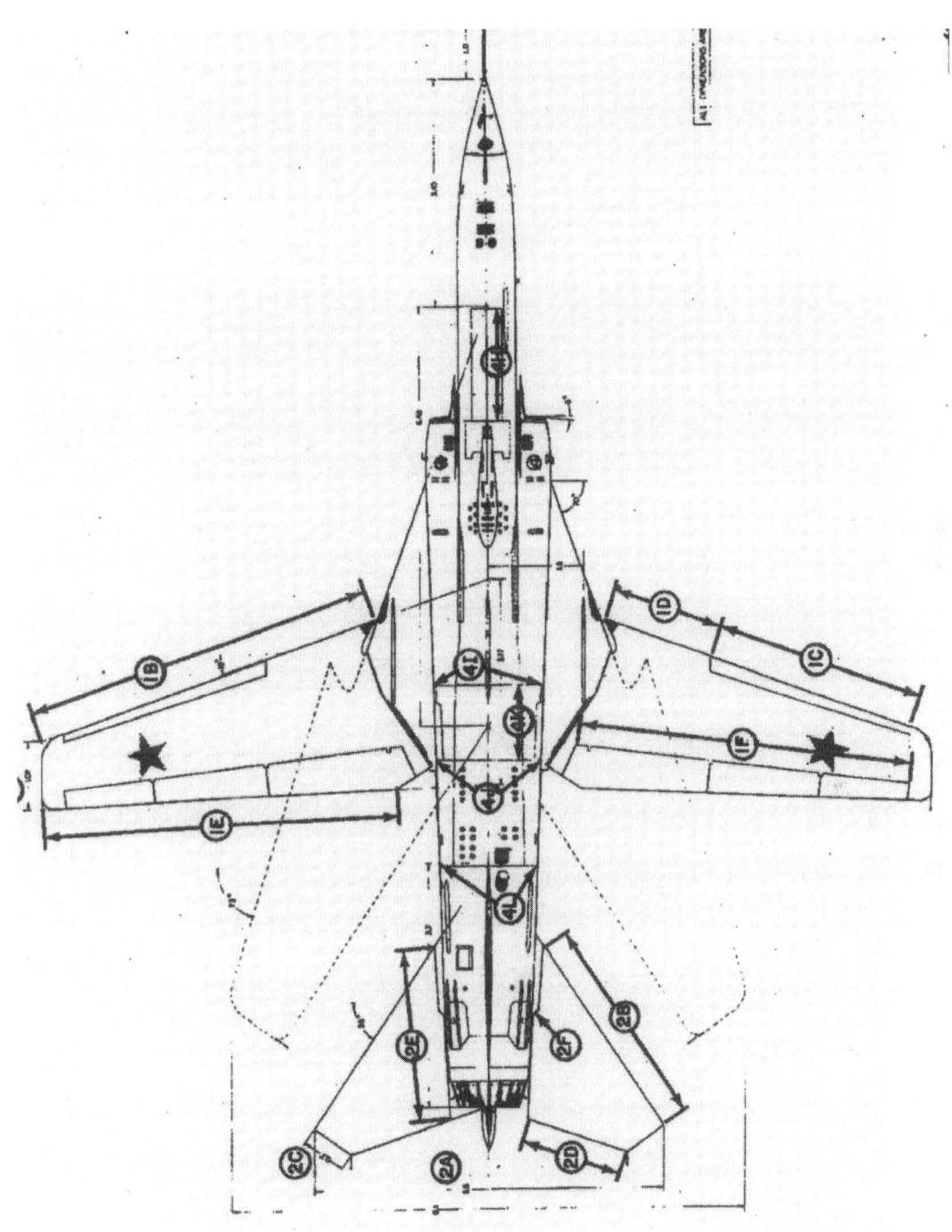

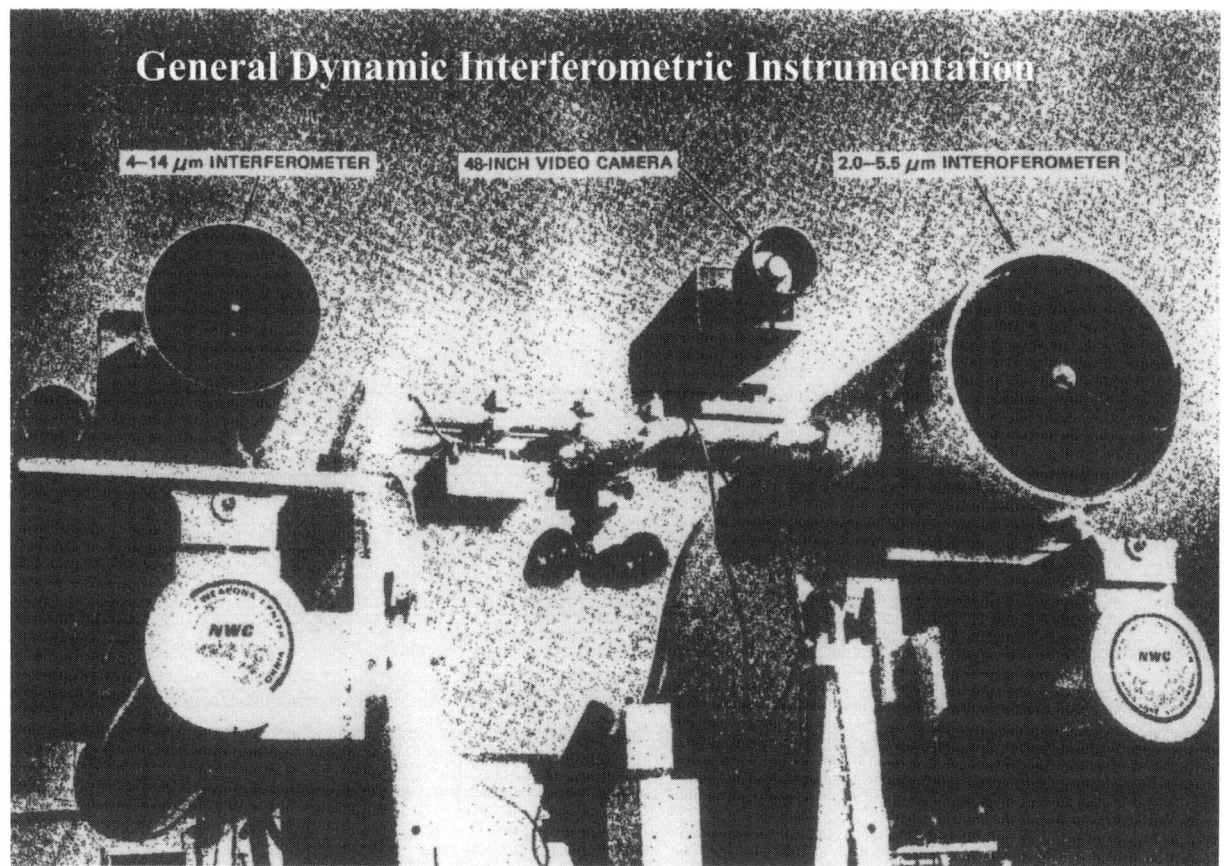

The R29-300, used in the MiG-23MS, was the latest known operational S.K. Tumanskiy power plant. Although similar in design philosophy to previously exploited Tumanskiy engines, i.e., R11, R13, and R15, those power plant reflected a significant improvement in terms of airflow handling capacity, thrust/weight, turbine inlet temperature, engine control logic and acceleration characteristics. Additionally, the engine used a unique variable turbine cooling scheme, which permitted essentially turning off the cool air when not required, thus improving subsonic cruise specific fuel consumption. The engine operated due the flight test program for approximately 87 hours with no major propulsion problems or pilot complaints. This engine reflected a continuation of Soviet air breathing propulsion design philosophy, with the major driver on engine cycle selection and design being cost, simplicity, and operational capability with little emphasis on time between overhaul, manufacturing man-hours, or on range. In the structures area, the MiG-12 employed essentially the same materials and manufacturing techniques found on earlier Soviet aircraft.

The FLOGGER used a RSBN-6s short-range navigation and landing system airborne orthodromic navigation system in conjunction with ground-based transmitter to perform tactical navigation and instrument landing. The pilot could select a pre-programmed destination — one of four ground stations or one of three route waypoints (points defined with respect to the ground stations) — or home base.

Its ALMAZ-23 (JAY BIRD) airborne intercept (AI) radar provided target search, acquisition, and grand and air-to-air missile (AAM) target illumination. JAY BIRD operated in the J-band, which provided a sharper beam, better angular resolution, and better control over side lobes. The radar maintained automatic tracking of the target under most target jamming conditions. If noise jamming denied range tracking, the radar could still track-on-jam (TOJ) or derive angle data by tracking the jamming source. Range data supplied via data link provided the remaining information necessary for missile launch. The SIRENA-3M radar warning system provided passive warning of incoming radar signals to the pilot by a lighted display and a tone in the headphones.

A rather significant finding of HAVE PAD exploitation was the use of V95 aluminum alloy for the primary wing skins, an alloy that had caused the US serious metal problems through the 1960s.

The well laid out cockpit of the FLOGGER had many human factors apparent. It was also an obvious Mikoyan design very similar to the later FISHBED MiG-21 and FOXBAT MiG-25. The canopy of the FLOGGER was a pneumatically operated clamshell that hinged aft for access. The emergency air operated jettison system on the FLOGGER was independent with a type KM-1M ejection seat system designed to operate from 70 - 540 KIAS at altitudes from ground level to over 18,000 m. The seat was catapult and rocket powered using four pyrotechnic devices and numerous precooked springs for sequencing and actuation. The pilot initiated ejection by gripping one or both ejection handles, pushing the locking lever(s) to unlock a ball lock securing the handle to the seat pan and pulling the handles(s) upward. During the initial vertical movement, a lanyard-actuated spring ensnared the pilot's legs to prevent flailing. After the seat cleared the aircraft, a three-chute stabilization, speed-retardation, and personnel-descent system sequenced the chute deployment and release according to the ejection speed and altitude.

The Soviet VKK-4 and VKK-6 partial-pressure suits were similar to the USAF model MC-4 suit. A capstan system provided anti-G protection extending along both legs from approximately the bottom rib down to the ankle region. The FLOGGER pilot used one of two types of standard flying helmets, depending on his mission — a crash helmet for subsonic flights at low to medium altitudes and a pressure helmet for high altitude and/or high subsonic flights.

The NWC conducted the ground-based IR measurements of the FLOGGER, however other commitments negated the use of some of the EG&G Special Project's radar test facilities, limiting this exploitation to the recorded IR spectral data recorded by contractor General Dynamics during the ground-to-air tests.

One aircraft sortie was devoted to radar signature measurements obtaining I and G band aspect measurements of the HAVE PAD vehicle.

Epilogue

It was probable even today that the US and its ally aviators routinely test their skills against their adversary's latest plane available. Pilots, including the famous astronaut Lt. Gen. Thomas Stafford of Weatherford, Oklahoma, continued to fly the Nevada skies long after the HAVE DOUGHNUT, HAVE Drill, and HAVE FERRY programs. Published speculation held that at one time or another since the first MiG arrived at Groom Lake in 1968, the US has flown in the Nevada desert every warplane in the Soviet arsenal, including at least two MiG-23 Floggers confirmed by actual sightings. Black programs began by HAVE DOUGHNUT gaining intelligence by flying additional Soviet-built fighters in the US reportedly peaked in the late 1980s and ended in the 1990s. The training of US pilots in the ways of their adversaries did not cease.

On April 26, 1984, Lt Gen Robert M 'Bobby' Bond who had flown the 'HAVE' MiG 15 years earlier-died in the crash of a MiG-23 Flogger on the Nellis range, one of the few occasions on which the USAF could not cover up post-1969 flights of the Soviet MiGs. Sadly, it was to have been the general's farewell flight, marking the end of a fine career. The high profile Bond mishap was a MiG accident the US Air Force could ill afford. Politics, rather than real security concerns accounted for most of the secrecy. Obviously, the Soviets knew what was going on, but the tests of these aircraft were kept secret, primarily to protect the identity (or identities) of the country (countries) providing the MiGs flown in Nevada.

Ironically, the world had been flooded with the latest in Soviet aircraft since the end of the Cold War. Virtually every aviation museum in the United States had on display the infamous MiG 21 plus most of the planes built afterwards. Nellis AFB in Nevada currently hosts a threat museum affectingly referred to as the petting zoo because of the museum's popularly among civilian tour groups wanting to have their photos taken in the cockpit of a MiG 29. NAS Fallon displays one of the finest collections of MiG planes that one can imagine. Almost every aviation museum in the United States now has Soviet MiGs on display. The American appetite for displaying the Soviet MiG is testimony to the lasting role the Soviet planes in American military aviation.

This current appetite to learn more about these mysterious planes started with the CIA's Special Projects team of inspired specialists who, with time on their hands between missions, the curiosity, the knowledge, and equipment to do so, experimented with learning the radar cross-section signatures of just about any plane available, including the MiGs. Twenty-four radar intercept sorties accomplished by the F-4 to evaluate the MiG-21 radar signature characteristics. The smallest radar cross section occurred from the head-on aspect, and average detection range was 20 nautical miles, Lock-on averaged 15 nautical miles from the head-on aspect. From a tail-on aspect, ranges increased to 25 nautical miles and 17 NM respectively. From abeam, or 90-degree aspect, the range for acquisition and lock-on increased to 35 and 28 nautical miles, respectively. Target altitude was determined to be a consideration only because ground clutter at lower altitude complicated the radar target recognition problem.

Comparison of APQ-109 (F-4D) radar detection ranges and those of the APQ-120 (F-4E) revealed that the APQ-109 acquired the MiG-21 at 510 percent greater range. The source of such a small deviation over a limited sampling was difficult to define. It recognized, however, that the increased beam width of the APQ-120, necessitated by a redesigned radar reflector to be compatible with the internal gun of the F-4E, decreases the amount of power concentrated on target, therefore, expecting a slight degradation in range performance.

Neither the F-4D nor F-4E was currently equipped with a suitable low altitude or look down capability against airborne targets. It, therefore, became evident that both aircraft (and the F-4C as well) were vulnerable, from a radar detection standpoint.

Constant Peg and the Red Eagles

By 1970, the HAVE Drill program expanded where a few selected fleet F-4 crews were given the chance to fight the MiG. The most important result of Project HAVE Drill was that no Navy pilot who flew in the project defeated the MiG-17 FRESCO in the first engagement. The HAVE Drill dogfights were by invitation only. The other pilots based at Nellis Air Force Base were not to know about the US-operated MiG. To prevent any sightings, the Air Force closed the airspace above the Groom Lake portion of the Nellis Range. On aeronautical maps, the exercise area was marked in red ink, the forbidden zone known as "Red Square".

USAF Colonel Gail Peck, a Vietnam veteran F-4 pilot, who was dissatisfied with his service's fighter pilot training, devised the idea of a more realistic training program for the Air Force. After the war, he worked at the Department of Defense, where he heard about the HAVE DRILL and HAVE DOUGHNUT programs. He won the support of USAF General Hoyt S. Vandenberg, Jr., and launched "Constant Peg," named after Vandenberg's callsign, "Constant," and Peck's wife, Peg.

Foreign military sales of United States fighter aircraft to Indonesia and Egypt in the mid-1970s to replace the Soviet fighter aircraft allowed these nations to clandestinely transfer unneeded MiG-21 ultra modern MiG-23s aircraft to the United States for evaluation. Up to 25 of these Soviet aircraft made their way to Groom Lake and pilots assigned to Detachment 1, 57th FWW at Nellis sent to the facility for training as "Aggressor" pilots before reassignment to the aggressor training units at Clark AB, Philippines, RAF Alconbury, England, and Nellis AFB. However, by the mid-1970s, the fleet of Soviet aircraft grew at Groom Lake, crowding the facilities. The Soviet Union knew of the MiGs at Area 51 and monitored their activities, so the MiG needed another clandestine home.

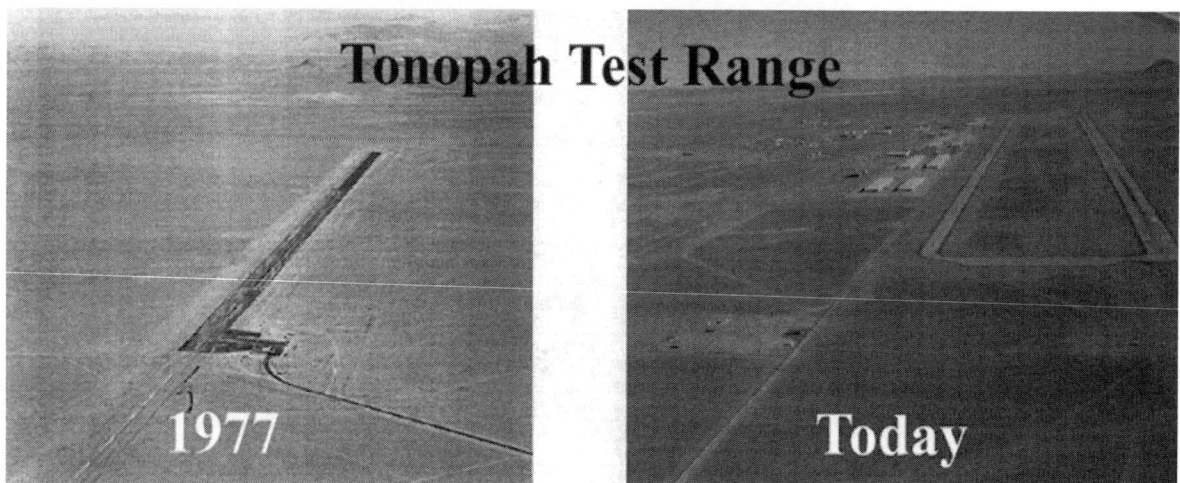

The several locations considered included Michael Army Airfield at the Dugway Proving Grounds in Utah, and the Gila Bend Air Force Auxiliary Field on the Goldwater Range in Arizona. However, the Tonopah Test Range Airport was only 70 miles to the southeast of Groom Lake and was on the controlled AEC Tonopah Test Range fitted the need for a new home. The AEC airport had the potential for improvement and expansion, with the only public land overlooking the base miles away. Although not as hidden as Groom Lake, the airport would be remote enough to operate the Soviet aircraft. In fact, the security surrounding the Tonopah Test Range was so effective that no one publicly reported the new base as an Air Force military airfield until 1985. On 1 April 1977, Air Force reassigned the 4477th TEF to Tonopah. In December 1977, Edwards AFB formed the 6513th Test Squadron ("Red Hats") to perform technical evaluations of these aircraft.

The product of Project Constant Peg, the 4477th Test and Evaluation Squadron (4477 TES), based at a remote airfield at Tonopah Test Range in the desolate desert north of Las Vegas, became a squadron in the United States Air Force under the TAC. There, they trained USAF pilots and weapon systems officers, and USN and USMC Naval Aviators and Naval Flight Officers to better fight the aircraft of the Soviet Union. Some 69 pilots, nicknamed Bandits, served in the squadron between 1979 and 1988, flying MiG-17s, MiG-21s, and MiG-23s.

The author cannot reveal the actual number or types of aircraft involved, where they came from, or the complete history of the program. As revealed through declassification, the activities of the 4477th Test and Evaluation Squadron brought about a fundamental change in United States Air Force, Navy, and Marine Corps air combat tactics. They revitalized the near forgotten art of dogfighting. The squadron's knowledge gained from testing the aircraft the squadron flew reflected in the success of United States air operations during the Vietnam War, as well as the founding of the Air Force's Red Flag program and the United States Navy's TOPGUN School.

In May 1973, Project Have Idea was formed which took over from the older HAVE DOUGHNUT and HAVE Drill projects and the project was transferred from the Area 51 facility to the Tonopah Test Range Airport, Nevada. At Tonopah, testing of foreign technology aircraft continued and expanded throughout the 1970s and 1980s.

By the late 1970s, United States MiG operations were undergoing another change. In the late 1960s, the MiG-17 and MiG-21F were still frontline aircraft. A decade later, later-model MiG-21s and new aircraft, such as the MiG-23, had superseded them. Fortunately, a new source of supply of Soviet aircraft became available, Egypt. In the mid-1970s, strained relations between Egypt and the Soviet Union caused Egypt to order out its Soviet advisers. The Soviets had provided the Egyptian air force with MiG since the mid-1950s. Now, with their traditional source out of the picture, the Egyptians began looking west. They turned to United States companies for parts to support their late-model MiG-21s and MiG-23s. Very soon, they entered into a deal where, according to one account, Egyptian president Anwar Sadat gave the United States two MiG-23 fighter-bombers disassembled and shipped from Egypt to Edwards Air Force Base. They then transferred initially to Groom Lake for reassembly and study.

In 1987, the US Air Force bought 12 new Shenyang F-7Bs from China for use in the Constant Peg program. At the same time, it retired the remaining MiG-21F-13 FISHBEDs acquired from Indonesia.

The United States operated MiG received special designations. There was the practical problem of what to call the aircraft. To solve this, the US gave them numbers in the Century Series. The MiG-21s and Shenyang F-7Bs were called the "YF-110" (the original designation for the USAF F-4C), while the MiG-23s were called the "YF-113."

The focus of AFSC limited the use of the fighter as a tool with which to train the front line tactical fighter pilots. AFSC recruited its pilots from the AFFTC at Edwards Air Force Base, California, who were usually graduates from the Air Force Test Pilot School at either Edwards or the Naval Test Pilot School at NAS Patuxent River, Maryland. TAC selected its pilots primarily from the ranks of the Weapons School graduates at Nellis AFB.

The 4477th began as the 4477th Test and Evaluation Flight (4477 TEF), which began 17 July 1979. The name was later changed to the 4477th Test and Evaluation Squadron (4477 TES) in 1980. The 4477th began with three MiG: two MiG-17Fs and a MiG-21 loaned by Israel, who had captured them from the Syrian Air Force and Iraqi Air Force. Later, it added MiG-21s from the Indonesian Air Force.

The Air Force collected the aircraft at the Department of Energy's Tonopah Test Range, where the squadron flew the MiG-17s until 1982, but mostly MiG-21s and MiG-23s.

Two pilots of the 4477th died flying the Soviet planes. The pilots had no manuals for the aircraft, although some tried to write one, nor was there a consistent supply of spare parts, which the squadron refurbished or manufactured at high cost.

On 23 August 1979, a pilot lost control of the squadron's MiG-17F, USAF serial 002, the HAVE FERRY MiG from Area 51. US Navy Lieutenant M. Hugh Brown, 31, of the US Navy's Test and Evaluation Squadron FOUR (VX-4), "Bandit 12," originally of Roanoke, Virginia, entered a spin while dogfighting a US Navy F-5. Brown recovered, but entered a second irrecoverable spin too low to eject. The plane hit the ground at a steep angle near the Tonopah Test Range airfield boundary, killing the pilot instantly.

On 21 October 1982, USAF Captain Mark Postai crashed with a MiG-23.

On 26 April 1984, USAF Lieutenant General Robert M. "Bobby" Bond, then vice commander of AFSC, died attempting to eject after losing control of his MiG-23 while supersonic. A few hours after the crash, Headquarters, AFSC, at Andrews Air Force Base, Maryland issued a brief statement: "Lt. Gen. Robert M. Bond, vice commander, AFSC, killed today in an accident while flying in an Air Force specially modified test aircraft". Three-star generals do not generally fly test flights, so Bond's death attracted press interest. The fact that the Air Force also refused to identify the type of plane also raised questions. Early reports claimed he had been flying "a super-secret Stealth fighter prototype." The death of a three-star general led the Air Force to reveal that it was flying Soviet aircraft. Named in his honor are a boulevard cut-off between Eglin Air Force Base (main base) and Hurlburt Field in Florida.

Much like Area 51 at the time, the 4477th Test and Evaluation Squadron facilities at TTR were spartan, mostly doublewide trailers with the roofs weighed down by tires to prevent them blowing off in high desert winds.

The 4477th TEF, a full-fledged squadron and the operation developed realistic combat training operations featuring adversary tactics, dissimilar air combat training, and electronic warfare, acquiring over the years, more acquired acquiring approximately two dozen including ultra modern MiG-23s (designated YF-113s) from various sources that remain classified today. These include three Cuban pilots who brought their MiG to Florida, and a number of Chinese made MiG purchased outright from China via the front company Combat Core Certification Professionals Company (CCCP). Three Syrians flew their MiG-23 and MiG-29s to Turkey in 1988 followed by Soviet Captain Alexander Zuyev in 1989.

The Red Eagles contended with un-maintainable planes with no spare parts or technical manuals available. Ground crews resorted to reverse engineering aircraft components and manufacturing them from raw materials. Two US pilots lost their lives in catastrophic crashes. However, they, along with the Red Hats at Area 51, did learn the strengths and weakness of the Warsaw Pact's fast movers, information that proved to be invaluable to NATO fliers in any sort of East/West engagement. American pilots proved this in air-to-air combat every time they encountered these Soviet Cold War era aircraft in Iraq, Libya, Angola, and elsewhere.

As remembered by Gail Peck, Historian, TAC created the 4477th Test and Evaluation Flight (TEF) to host the training program chartered by M/Gen Hoyt S "Sandy" Vandenberg in s program called project CONSTANT PEG. CONSTANT was the "call sign" of M/Gen

Vandenberg and PEG was the wife of Major Gaillard (Gail) Peck who initiated the program while working for M/Gen Vandenberg at the Pentagon.

The flight of CONSTANT PEG trained USAF, US Navy and US Marine Corps combat fighter aircrews on the best ways to fight and win when encountering MiG aircraft in aerial combat. The 4477th TEF "stood up" at Nellis AFB, NV on 1 Apr 1977 under the command of Lt Col Glenn Frick, transitioned to the flight base at the Tonopah Test Range airfield under the command of Lt Col Gaillard Peck in Jul 1979 and terminated operations at Tonopah in Mar 1988 under the command of Lt Col Michael Scott. During these intervening years Lt Cols Earl Henderson, Tom Gibbs, George Gennin, Phil White, and Jack Manclark also commanded the unit.

During the time of active MiG operations, the 4477 TEF name changed to the 4477 Test and Evaluation Squadron (TES). Shortly after the termination of flight operations the squadron "stood down," the Air Force salvaged and redistributed the equipment to other users and reassigned all personnel.

During the period of operations, the 4477th TEF/TES flew over 15,000 MiG sorties and trained almost 6,000 Air Force, Navy, and Marine Corps aircrews to fight the MiG-17, MiG-21 and MiG-23 and win.

Lt Col Glenn Frick hired SMSgt Bobby Ellis to be the Chief of Maintenance. Bobby's task included hiring necessary personnel and supervising the management and restoration of the aircraft made available to the 4477th TEF.

Glenn Frick and his operations officer, then Maj. Ron Iverson (Lt Gen ret), hired the pilots that were to fly the aircraft.

Some MiG flying was ongoing under the cover of test operations. The birth of CONSTANT PEG marked the end of the "test charade" and signaled the beginning of MiG operations in earnest that were wholly dedicated to training American airmen in how to engage and beat the MiG aircraft in close-in aerial combat (dog fighting).

Gail Peck took command from Glenn Frick on 1 Oct 1978 and embarked upon the task of supervising the completion of all aspects of "standing up" the 4477th TEF including the completion of the airfield, the restoration of the MiG aircraft and the initiation of flight operations. When operations began at Tonopah Test Range in July 1979, the manpower consisted of 29 officers and enlisted personnel including three pilots that were US Navy aviators. Throughout the entire period, the 4477th was a Joint operation with USAF and US Navy and/or

Marine Corps pilots assigned. Initially, one civilian secretary supported the operations at her workstation at Nellis AFB, NV.

The initial equipage at Tonopah in Jul 1979 included eight aircraft: two MiG-17s, NATO FRESCO C aircraft and 6 MiG-21 NATO Fishbed C/E aircraft. The only vehicle initially assigned was a Kenworth tractor truck capable of 18-wheel trailer operations.

Through the good work of Mr. Matt Foley (RIP Jan 2013 at age 92), an AF intelligence officer, the force structure of RED EAGLE aircraft grew reaching a maximum of 27 flyable MiG airframes in 1985. Not all were flyable at the same time, as the maintainers moved certain critical parts from one flyable aircraft to make another aircraft flyable. The fleet was augmented by five T-38s that were used for MiG "chase" when new pilots were flying the single seat MiG for the first few times. None of the MiG was two-seaters and there were no simulators!

While initially RED EAGLE MiG pilots also flew mission aircraft at Nellis such as Aggressor F-5Es or Weapons School F-4Es, F-16s and F-15s, RED EAGLE pilots eventually evolved into logging time and taking check rides for their proficiency requirements in the assigned T-38s. As an added benefit, the T-38s enabled the unit to fly VIP orientations and provide incentive flights for the people assigned to the TTR (both MiG and F-117 personnel).

The unit also had a small fleet of civilian light twin aircraft (Cessna 404s later replaced by Mitsubishi MU-2s) that were used for transportation, mostly between Nellis and TTR. RED EAGLE pilots flew the light twins. The Air Force eventually phased out the light twins in favor of AF C-12s flown by non-project military pilots.

The MiG-17s were retired in 1981 after an engine problem prompted Capt Mark Postai to crash land in the desert. He stated after escaping uninjured that it was a very rough ride and something he would not likely do again.

The drawdown of the MiG-17 fleet corresponded with the introduction of the MiG-23 Flogger and the acquisition of additional MiG-21s. At the peak, the 4477th had 17 MiG-21s and 10 MiG-23s. An increased number of personnel assigned along with the acquisition of, and in some cases, the manufacture of additional surface vehicles accompanied the buildup of the MiG fleet that the RED EAGLES acquired from Defense Reutilization Management Offices (DRMO) or salvage yards throughout the West.

Three RED EAGLES died during the MiG program. LT (USN) Hugh Brown died in the crash of the HAVE FERRY MiG-17 and Capt Mark Postai in the crash of a MiG-23. TSgt Rey Hernandez, a fuels specialist, died because of injuries from an industrial incident relating to the repair of a fuel cell in a RED EAGLE T-38 plane.

Eventually the cost of keeping the MiG in the air caught up to the program. By the late 1980s, the decline of communist rule in the Soviet Union rendered Constant Peg anachronistic. The Air Force ended the program in March 1988, but did not deactivate the squadron until July 1990.

During the 2006 declassification of the Constant Peg program, the USAF held a series of press conferences about the former top-secret US MiG. The US MiG flew more than 15,000 sorties and nearly 7,000 aircrew flew in training against dissimilar aggressors in the Nevada desert between 1980 and the end of the program in 1988.

Red Hats

On 1 April 1977, the Air Force reassigned the 4477th TEF to Tonopah. In December 1977, the 6513th Test Squadron ("Red Hats") formed at Edwards AFB to perform technical evaluations of the Soviet MiG aircraft at Area 51 and flown by the Red Eagles at the Tonopah Test Range in Nevada.

Testing of foreign technology aircraft began at Area 51 in 1968 had continued and expanded throughout the 1970s and 1980s with additional MiG-17, MiG-21, MiG-23, Su-7B, Su-22, and other aircraft undergoing intensive evaluations.

Project HAVE GLIB

Area 51 foreign materiel evaluation program HAVE GLIB involved testing Soviet tracking and missile control radar systems. A complex of actual and replica Soviet-type threat systems began to grow around "Slater Lake" (the pond, which had been named after the former Roadrunners commander), a mile northwest of the main base. They arranged them to simulate a Soviet-style air defense complex. The Air Force began funding improvements to Area 51 in 1977 under project SCORE EVENT.

Defections with MiG planes continued

Over the years, more and more aircraft added to the collection. On September 6, 1976, Lieutenant Viknotsor Belenko defected with his MiG-25 to Hakodate, Japan. The US interrogated and debriefed him for 5 months after his defection, and employed him as a consultant for several years thereafter. The MiG, disassembled and examined, returned to the USSR in thirty crates. Belenko brought with him the pilot's manual for the Foxbat, expecting to assist American pilots in evaluating and testing the aircraft. However, the Japanese government only allowed the US to examine the plane and do ground tests of the radar and engines. In July 1988, two Syrian pilots defected with their MiG-29s to Turkey followed by another Syrian pilot in April same year who landed his MiG-23ML in Turkey. In October 1989 Syrian pilot Abdel Bassem landed his MiG-23ML in was Israel.

The USAF continues a Foreign Materiel Acquisition/Exploitation program, although the extent of acquisitions and operations of that program is most likely classified and unavailable. In March 1991, in the aftermath of the 1991 Gulf War, an exploitation team from the Joint Captured Materiel Exploitation Center arrived at Jalibah Southeast Air Base in Iraq. They returned with a MiG-29 nose, providing Air Force intelligence personnel with a Slot Back I radar and the Fulcrum's infrared search and tracking system. Later in the decade, Air Force intelligence personnel were able to acquire more complete versions of the MiG-29, the result of spending money rather than fighting a war. In October 1997, the US purchased 21 fighter aircraft from the Republic of Moldova—including the MiG-29UB. According to the National Air and Space Intelligence Center, after "undergoing years of study" and employing "all the [center's foreign materiel exploitation] resources," the MiG-29UB was displayed in front of NASIC headquarters at Wright Patterson AFB, Ohio.

An Iraqi Air Force MiG-25 Foxbat found buried under the sand at Al-Taqaddum Air Base, Iraq, 2003

In 1997, the United States purchased 21 Moldovan aircraft for evaluation and analysis, under the Cooperative Threat Reduction accord. Fourteen were MiG-29Ss equipped with an active radar jammer in its spine and capable nuclear weapon armament. Part of the United States' motive to purchase these aircraft was to prevent them from being sold to "rogue states," especially Iran. In late 1997, the MiG arrived at the National Air and Space Intelligence Center

(NASIC) at Wright-Patterson Air Force Base near Dayton, Ohio, though many of the former Moldovan MiG-29s were most likely scrapped.

In 2003, after the seizure of the Iraqi Air Force Al-Taqaddum Air Base, US troops found an advanced Russian MiG-25 Foxbat buried in the sand. US Air Force recovery exploitation teams dug the MiG out of a massive sand dune near the Al Taqqadum airfield by. The MiG was reportedly one of over two dozen Iraqi jets buried in the sand like hidden treasure and recovered later. Reportedly, not all the jets found at captured Iraqi Air Force bases were from the Gulf War era. The Russian-made MiG-25 Foxbat recovered was an advanced reconnaissance version never before seen in the West and was equipped with sophisticated electronic warfare devices. Air Force recovery exploitation teams had to use large earth-moving equipment to uncover the MiG, which was over 70 feet long and weighed 25 tons. The advanced electronic reconnaissance version found by the US Air Force is currently in service with the Russian air force.

Aircraft receiving US designations and fake serial numbers to identify them in DOD standard flight logs:

YF-110B MiG-21F-13 "FISHBED-C"
YF-110C A MiG-21 Chinese Chengdu J-7B (MiG-21F-13 A FISHBED variant)
YF-110D a MiG-21 "FISHBED" variant NATO
YF-112 Sukhoi Su-22 Fitter
YF-113A MiG-17F "FRESCO-C" used in HAVE DRILL program
YF-113B MiG-23BN "Flogger-F" NATO
YF-113C MiG-17F (actually a Chinese-built J-5) "FRESCO-C" used in HAVE PRIVILEGE program
YF-113E MiG-23MS "Flogger-E" NATO
YF-114C MiG-17F "FRESCO-C" (the one used in the HAVE FERRY program)
YF-114D MiG-17PF "FRESCO-D NATO: (Serial: 75-008)
YF-116 MiG-25 Foxbat
YF-118 MiG-29 Falcrum

MiGs in Nevada exposed

In April 1984, Lt. Gen. Robert M. Bond made two orientation flights in a Russian-built MiG-23 jet fighter. While making a high-speed run during his second flight, Bond lost control and crashed in Area 25 of the Nevada Test Site. He died while ejecting.

The 4477th flew its last MiG in 1990 and deactivated with the end of the Cold War. The 6513th which stopped flying MiG in the early eighties, inactivated in 1992. In 2006, the Constant Peg program was declassified and the USAF held a series of press conferences about the former top-secret US MiG. US MiG flew more than 15,000 sorties and nearly 7,000 aircrew flew in training against dissimilar aggressors in the Nevada desert between 1980 and the end of the program in 1988. These included the 'real' MiG as well as US made A4 and T-38/F5 aircraft visually modified to be 'fake' MiG. The Groom Lake facility still exists (unofficially, of course).

Other countries following the lead of the USAF and USN in aerial combat training

The Argentine Air Force trains its fighter pilots in the CEPAC (Course Combat Airman Standardization) course given at the IV Air Brigade in the town of "El Plumerillo" located in the province of Mendoza. The officers received from the Military Aviation School receive academic instruction and flight-training aircraft in the AT-63 Pampa advanced, and then integrate some of the operational squadrons of the Air Force.

The 410 Squadron of the Royal Canadian Air Force conducts an annual Fighter Weapons Instructor Course (FWIC) at CFB Cold Lake in Alberta. The course is three months in length and is specific to the CF-18 Hornet aircraft. There are eight students per course.

In Greece, the Hellenic Air Force built its own school in 1975 called the Tactical Weapons School based in Andravida Air Base. In 1983, Hellenic Air Force established the KE.A.T. (Meaning Air Tactics Centre in Greek). Now the Tactical Weapons School is part of the KE.A.T. Both, based in Andravida, act as an independent squadron of the Hellenic Air Force. Every year the best pilots from all the squadrons of the Hellenic Air Force train in modern air-to-air tactics, air-to-ground tactics, COMAO packages, and Electronic Warfare in KE.A.T. The pilots graduating are the best pilots in the Hellenic Air Force.

The Royal Netherlands Air Force has a Fighter Weapons Instructor Training (FWIT) with 323 Tactical Training, Evaluation & Standardization Squadron (TACTESS) at Leeuwarden Airbase. This training is a multi-national effort with Norway, Denmark, Belgium, and Portugal.

The Pakistan Air Force initially provided similar training via the fighter leader's school but the dedicated Combat Commanders School at PAF Base Mushaf replaced this in 1976. The Combat Commanders' School continues to fulfill its mission of training fighter pilots and air defense controllers. In order to keep pace with the ever-changing aerial threat and environment, CCS reviews its courses content continually, adding new study and flying phases involving EW and BVR threat along with counter-insurgency tactics and others augmented to maintain pace with current Air warfare trends.

Turkish Air Force has a similar air-to-air and air-to-ground simulated warfare conditions training program based on highly developed ACMI (Air Combat Maneuvering Instrumentation) system, named Anatolian Eagle. Each year several countries participate into operations including Belgium, France, Germany, Israel, Italy, Jordan, Netherlands, United Arab Emirates, United Kingdom, and United States of America.

Located in Konya 3rd. Main Jet Base, the main missions of the program included systematically testing and evaluating the fighters' combat readiness statuses and managing the tactical training progress. The program built a background and knowledge base in order to make research on tactical aeronautics, to make research to allow fighter elements of the Turkish Air Force Command. The program reaches the military goals in the shortest time and with minimum resource and effort, to support the definition of operational requirements and supply and R&D activities, to allocate training environment in order to fulfill the requirements of the Turkish Air Force Command, to support the tests of existing/developed/future weapon/aircraft systems.

In the United Kingdom, the Royal Air Force and Royal Navy also has a similar course specific to each aircraft type, known as the QWI (Qualified Weapons Instructor, pronounced Que-Why) Course. It is five months in length.

The Russian Air Force also has its own specialty course in modern air-to-air and air-to-ground combat based at Lipetsk Air Base called the 4th Center of Combat Application and Conversion of Frontline Aviation and under the command of General-Major Aleksander Kharchevsky. Lipetsk today has many of the old, current, and new Russian Air Force hardware, including the Sukhoi Su-34 and Yakovlev Yak-130. Once the T-50 PAK-FA fifth generation fighter come into service, the first ten series production copies go to Lipetsk for training of instructors. The new Sukhoi Su-35 will provide training for the new aircraft, as well at the T-50

WORLD WIDE FISHBED AOB
(S-Gp-1)

NATIONAL AIR FORCES	MODEL C/E	D/F	UNK	TOTAL
Bulgaria	22	14		36
Czechoslovakia	42	76	16	134
E Germany	76	141		217
Hungary	60	30		90
Poland	43	68		111
Rumania	42	10		52
Yugoslavia	36	20		56
Russia		717		717
Communist China	33		13	46
North Korea	1	10	11	22
North Vietnam			6	6
Indonesia	17			17
Cuba	33	27		60
Syria	1			1
Iraq	11	13		24
India			64	64
United Arab Republic		36	72	108
Afganistan			32	32
	417	1162	214	1793

Soviet Air Forces in European Communist Countries

E Germany		284		284
Hungary		111		111
Poland		111		111
		506		506

TOTAL FISHBED - WORLD WIDE   2299

United States Navy TOPGUN

Immediately following the HAVE DOUGHNUT, HAVE DRILL, and HAVE FERRY exploitation projects at Area 51, the United States Navy recognized the need to train its pilots against the MiG planes they faced in SEA, and against the tactics used against them.

In May 1968, the Navy published the "Ault Report," which concluded that the problem stemmed from inadequate air-crew training in air combat maneuvering (ACM). The F-8 Crusader community, who had been lobbying for an ACM training program ever since Rolling Thunder began, welcomed this. The Ault Report recommended establishment of an "Advanced Fighter Weapons School" to revive and disseminate community fighter expertise throughout the fleet. CNO Moorer concurred.

On March 3, 1969, the Naval Air Station Miramar north of San Diego established the United States Navy Weapons School. In 1996, the School merged into the United States Navy Strike Fighter Tactics Instructor program (SFTI program). Today, more popularly known as TOPGUN, the Naval Strike and Air Warfare Center at Naval Air Station Fallon, Nevada teaches fighter and strike tactics and techniques to selected Naval Aviators and Naval Flight Officers, who returned to their operating units as surrogate instructors.

The Navy formed the school using many F-8 pilots as instructors, and placed under the control of the VF-121 "Pacemakers" an F-4 Phantom-equipped Replacement Air Group (RAG) unit, borrowing aircraft from its parent unit and other Miramar-based units.

Based on the Naval pilots first experiences with the MiG in Nevada, the objective was to develop, refine and teach aerial dogfight tactics and techniques to selected fleet air crews to realistically replicate expected enemy aircraft and is widely used in air arms the world over. At that time, the predominant enemy aircraft were the Russian-built transonic MiG-17 'FRESCO' and the supersonic MiG-21 'FISHBED' that the Naval pilots were engaging in SEA and had recently engaged in Nevada.

The TOPGUN course chose air crews from front-line units, who upon graduating, returned to their parent fleet units to relay what they had learned to their fellow squadron mates—in essence becoming instructors themselves.

From the data provided the Navy from the HAVE DOUGHNUT and HAVE Drill tests, the newly formed United States Navy Fighter Weapons School (TOPGUN) at NAS Miramar, California changed the remainder of the Vietnam War, climbing the Navy kill to 8.33:1. In contrast, the Air Force rate improved only slightly to 2.83:1. The reason for this difference was

TOPGUN. The Navy (to include the Marine Corps) had revitalized its air combat training, while the Air Force had stayed stagnant. Most of the Navy MiG kills were by TOPGUN graduates. The Air Force, which had not implemented a similar training program, actually had its kill ratio worsen.

The success of the US Navy fighter crews led to TOPGUN becoming a separate, fully funded command in itself, with its own permanently assigned aviation, staffing, and infrastructural assets. Successful TOPGUN graduates scoring air-to-air kills over North Vietnam returned to instruct at TOPGUN. These included "Mugs" McKeown and Jack Ensch who flew against the MiGs in Nevada.

In 1996, NAS Miramar transferred to the Marine Corps at which time the Navy moved TOPGUN into the Naval Strike and Air Warfare Center (NSAWC) at NAS Fallon, Nevada where it continues to conduct an Adversary Training Course, flying with adversary aircrew much as it did during HAVE DOUGHNUT, HAVE DRILL, and HAVE FERRY. An admiral commands the Naval Strike and Air Warfare Center at NAS Fallon, the Navy center of excellence for naval strike and air warfare.

Exploitation at Area 51 developed Air Force and Navy training programs

The Navy Top Gun Fighter Weapons School and AF Red Flag training exercises turned the kill ratio around where by the war's end; the US Navy/Marine Corps recorded 16 jets lost to NVAF MiG, while shooting down 61 NVAF MiG-17, MiG-19, and MiG-21s. The USAF records show losing 67 jets to the NVAF. The NVAF in some writings might claim well over those figures. Be advised the NVAF MiG 17s & 19s were not all Soviet built aircraft; many of the MiG-17s were actually Chicom J5 versions of the Soviet MiG-17F & the MiG-19s were actually Chicom J6s (Chinese Communist). The Soviets did NOT supply the NVAF with MiG-19s. The Red Chinese did NOT supply the NVAF with MiG-21s. However, they both supplied the NVAF with MiG-17s (and J5s).

While various studies and reports offered many suggestions, the basic fact of the matter is that Projects HAVE DOUGHNUT, HAVE DRILL, and HAVE FERRY revealed the reason for the lopsided aerial losses being the USAF and USN having lost the art of dogfighting as a skill. In the USAF, for instance, the solution in the 1960s to an increasing accident rate in the McDonnell Douglas F-4 Phantom II was simply to ban air combat maneuvering training (ACM). The accident rate fell, but legions of Air Force Phantom drivers entered the skies of Vietnam with little experience in knowing what their aircraft could and could not do in a dogfight with North Vietnamese MiG.

The United States Air Force did not react to what it learned during HAVE DOUGHNUT, HAVE DRILL, and HAVE FERRY as quickly as did the Navy. Consequently, the Air Force air crews fighting the war in SEA continued to rate poorly with unacceptable performance of US Air Force fighter pilots and weapon systems officers (WSO) in air combat maneuvering (ACM) (air-to-air combat) during the Vietnam War in comparison to previous wars. Air combat over North Vietnam between 1965 and 1973 led to an overall exchange ratio (ratio of enemy aircraft shot down to the number of own aircraft lost to enemy fighters) of 2.2:1 (for a period of time in June and July 1972 during Operation Linebacker the ratio was less than 1:1). Among the several factors resulting in this disparity was a lack of realistic ACM training as learned when the Air Force encountered the MiG at Area 51 in Nevada.

It was not until 1975 that an Air Force analysis known as Project Red Baron II showed that a pilot's chances of survival in combat dramatically increased after he had completed 10 combat missions. As a result, the United States Air Force created Red Flag to offer USAF pilots and weapon systems officers the opportunity to fly 10 realistically simulated combat missions in a safe training environment with measurable results. Many US aircrews had also fallen victim to SAMs during the Vietnam War and Red Flag exercises provided pilots and WSOs experience in this regime as well.

The concept of Colonel Richard "Moody" Suter became the driving force in Red Flag's implementation, persuading the then-TAC commander, General Robert J. Dixon, to adopt the program. At Nellis, Suter was well known and well liked. The first Red Flag exercise came off on Gen Dixon's schedule in November 1975. On 1 March 1976, the USAF chartered the 4440th Tactical Fighter Training Group (Red Flag) with Col P.J. White as the first commander, Lt Col Marty Mahrt as vice commander, and Lt Col David Burner as Director of Operations. This small crew under Col White's leadership undertook the task of firmly establishing the program.

They selected the "aggressor squadrons," the opponents who flew against the pilots undergoing training, from the top fighter pilots in the US Air Force. They trained these pilots to fly according to the tactical doctrines of the Soviet Union and other enemies of the period, in order to better simulate what then-TAC, as well as USAFE, PACAF and other NATO pilots and WSOs would likely encounter in real combat against a Soviet, Warsaw Pact, or a Soviet-proxy adversary. The Air Force originally equipped the aggressors with readily available T-38 Talon aircraft to simulate the MiG-21, the T-38 being similar in terms of size and performance. F-5 Tiger II fighters, painted in color schemes commonly found on Soviet aircraft, added shortly thereafter became the mainstay until introduction of the F-16.

In 1975, the United States Air Force conducted its first advanced aerial combat training exercise at Nellis Air Force Base, Nevada known as Red Flag. Since 1975, aircrews from the United States Air Force (USAF), United States Navy (USN), United States Marine Corps (USMC), United States Army (USA) and numerous NATO or other allied nations' air forces have taken part in one of several Red Flag exercises held during the year, each of which is two weeks in duration.

Under the aegis of the United States Air Force Warfare Center (USAFWC) at Nellis, the Red Flag exercises continue today. Conducted in four to six cycles a year, the 414th Combat Training Squadron (414 CTS) of the 57th Wing (57 WG) trains pilots and other flight crew members from the US, NATO and other allied countries for real air combat situations. This includes the use of "enemy" hardware and live ammunition for bombing exercises within the adjacent Nevada Test and Training Range (NTTR). At NTTR, the 414th Combat Training Squadron (Red Flag) mission maximizes the combat readiness and survivability of participants by providing a realistic training environment and a pre-flight and post-flight training within the Nellis Range Complex located northwest of Las Vegas and covering an area of 60 nautical miles.

In a typical Red Flag exercise, Blue Forces (friendly) engage Red Forces (hostile) in realistic combat situations.

Blue Forces are made up of units from the Air Combat Command (ACC), Air Mobility Command (AMC), Air Force Global Strike Command (AFGSC), Air Force Special Operations Command (AFSOC), United States Air Forces Europe (USAFE), Pacific Air Forces (PACAF), Air National Guard (ANG), Air Force Reserve Command (AFRC), and Air Force Space Command (AFSPC), aviation units of the US Navy, US Marine Corps and US Army, the Royal Air Force, Royal Canadian Air Force, and Royal Australian Air Force, as well as other allied air forces and fleet air arms. A Blue Forces commander, who coordinates the units in an "employment plan" scheme of operation, leads them.

Red Forces (adversary forces) are composed of the 57th Wing's 57th Adversary Tactics Group (57 ATG), flying F-16s from the 64th Aggressor Squadron (64 AGRS) and F-15s from the 65th Aggressor Squadron (65 AGRS) to provide realistic air threats through the emulation of opposition tactics. Other US Air Force, US Navy, and US Marine Corps units flying in concert with the 507th Air Defense Aggressor Squadron's (507 ADAS) electronic ground defenses and communications, and radar jamming equipment also augment the Red Forces. The 527th Space Aggressor Squadron (527 SAS), an Active Duty unit, and the 26th Space Aggressor Squadron (26 SAS), an Air Force Reserve Command unit, also provide GPS jamming. Additionally, the Red Force command and control organization simulates a realistic enemy integrated air defense system (IADS).

A key element of Red Flag operations is the Red Flag Measurement and Debriefing System (RFMDS), a computer hardware and software network providing real-time monitoring, post-mission reconstruction of maneuvers and tactics, participant pairings and integration of range targets and simulated threats. Blue Force commanders objectively assess mission effectiveness and validate lessons learned from data provided by the RFMDS.

A typical flag exercise year includes ten Green Flags (a close air support (CAS) exercise with the US Army), one Canadian Maple Flag (operated by the Royal Canadian Air Force), and four Red Flags. Each Red Flag exercise normally involves a variety of fighter interdiction, attack/strike, air superiority, enemy air defense suppression, airlift, air refueling, and reconnaissance missions. In a 12-month period, more than 500 aircraft fly more than 20,000 sorties, while training more than 5,000 aircrews and 14,000 support and maintenance personnel.

Before a "flag" begins, the Red Flag staff conducts a planning conference where unit representatives and planning staff members develop the size and scope of their participation. All aspects of the exercise, including billeting of personnel, transportation to Nellis AFB, range coordination, ordnance/munitions scheduling, and development of training scenarios, are designed to be as realistic as possible, fully exercising each participating unit's capabilities and objectives.

Today, the 414th Combat Training Squadron (414 CTS) is the unit currently tasked with running Red Flag exercises, while the 64th Aggressor Squadron (64 AGRS) and the 65th Aggressor Squadron (65 AGRS) also based at Nellis AFB use F-16 and F-15 aircraft Flanker painted in the various camouflage schemes of potential adversaries to emulate, respectively, the MiG-29 Fulcrum and Su-30.

The US Air Force's Red Flag approach differs from that initially employed during and after the Vietnam War by the United States Navy to improve fighter aircrew performance. Rather than a large, multi-squadron exercise, the Navy established the United States Navy Fighter Weapons School (more widely known as TOPGUN) in 1969 at the former NAS Miramar, California. TOPGUN "trained the trainers," with Navy and Marine Corps squadrons in the Fleet selecting their best fighter, strike fighter, and Marine fighter/attack aircrews for training when back at their home stations between overseas deployments. TOPGUN graduates then returned to their Fleet squadrons prior to their next overseas deployment to share lessons learned with their fellow Naval Aviators and Naval Flight Officers.

Navy and Marine Corps adversary squadrons were also later established at Master Jet Bases. These bases include NAS Miramar (now MCAS Miramar), NAS Oceana, NAS Lemoore, MCAS Yuma, MCAS Beaufort, the former NAS Cecil Field, the former MCAS El Toro. The bases also include fleet air training bases such as NAS Fallon, NAS Key West, and the former

NS Roosevelt Roads and forward deployed bases such as the former NAS Cubi Point, for Fleet squadrons to conduct dissimilar air combat training (DACT) as part of unit level training (ULT).

These adversary squadrons initially flew the A-4 Skyhawk, with the Navy and Marine Corps later adding the T-38 Talon and F-5E/F to its adversary lineup and briefly included the F-21 Kfir. Other naval adversary aircraft have included specially built F-16Ns, the F-14, and the F/A-18. Today, Carrier Air Wing level training, analogous to the USAF Red Flag program, is conducted at NAS Fallon, where the Naval Strike and Air Warfare Center (NSAWC), of which TOPGUN in now part, operates dissimilar adversary aircraft (F-16 and F/A-18), while a collocated Naval Air Reserve squadron, Fighter Composite Squadron 13 (VFC-13), flies the F-5E and F-5F.

The United States Marine Corps (USMC) also conducts Weapons and Tactics Instructor (WTI) exercises at Marine Corps Air Station Yuma twice a year as part of the WTI course conducted by Marine Air Weapons and Tactics Squadron ONE (MAWTS-1) and uses locally based Marine Fighter Training Squadron 401 (VMFT-401), a Marine Air Reserve squadron and the only USMC adversary squadron. Originally equipped with the F-21 Kfir, VMFT-401 now operates the F-5E and F-5F.

In 2009, the 416th Flight Test Squadron (416 FLTS) from Edwards AFB, California, also participated in Red Flag, the first time an Air Force Material Command (AFMC) unit had been part of the program.

In the Southeast Asian conflict, however, that exchange ratio fell to less than 1-to-1 during a period in the spring of 1972. Today's Red Flag over a single year will involve as many as 250 different units and 750 aircraft of many different types. About 11,000 aircrew and squadron personnel will amass more than 12,000 sorties and 21,000 flight hours in the course of the year. One major milestone in that history, without question, was the stunning performance of American airmen in the Gulf War of 1991. It was the first war to highlight the results of Red Flag, and it produced a curious tribute when an Air Force pilot returning from a combat mission over Iraq remarked, "It was almost as intense as Red Flag."

Participating countries

Only countries considered friendly towards the United States take part in Red Flag exercises. So far, the countries to have participated in these exercises are:
- Australia
- Belgium
- Brazil (July 1998, August 2008 and February 2013)
- Canada
- Chile (July 1998)
- Colombia (July 2012)
- Denmark
- Egypt
- France
- Germany
- Greece (October 2008)
- India
- Israel
- Italy

- Japan
- New Zealand
- Netherlands
- Norway
- Pakistan (2010)
- Poland (June 2012)
- Portugal (March 2000)
- Sweden
- Singapore
- Saudi Arabia
- South Korea
- Spain
- Thailand
- Turkey
- United Arab Emirates
- United Kingdom
- Venezuela

References

- Foreign Technology Division, US Air Force Systems Command FTD-CS-09-5-67, Secret Publication of 18 July 1968, "FRESCO (MiG-17) Weapon System "
- NAVAIR-01-245FDB-1 "NATOPS Flight Manual F-4B Aircraft" of 15 Feb 1969
- NAVAIR 01-245FDB-1A "Supplement to NATOPS Flight Manual F-4B Aircraft" of 15 Feb 1969
- NAVAIR 01-245FDD-1 "NATOPS Flight Manual F-4J Aircraft" of 15 Jun 1967
- NAVAIR 01-245FDD-1A "Supplement to NATOPS Flight Manual F-4J Aircraft" of 1 Jun 1968
- NAVAIR "01-45HHE-1 NATOPS Flight Manual F-8H Aircraft" of 1 Could 1967
- NAVAIR 01-45HHE-1A "Supplement to NATOPS flight Manual F-8H Aircraft" of 1 Could 1967
- NAVAIR 01-45HHF-1 "NATOPS Flight Manual F-8J Aircraft" of 15 Apr 1968
- NAVAIR 01-45HHF-1A "Supplement to NATOPS Flight Manual F-8J Aircraft" of 15 April 1968
- NAVAIR 01-40ABE-1, "NATOPS Flight Manual A-4/Ta-4 " of 15 Feb 1967
- NAVAIR 01-85ADA-1, "NATOPS Flight Manual A-6A " of 1 Dec 1967
- NAVAIR 01-45AAA-1 "NATOPS Flight Manual A-7 " or 1 Jan 1968
- NAVAIR 01-245FDB-1T Tactical Manual for F-4B/J Aircraft of 1 Mar 1969
- NAVAIR 01-245FDB-1T Supplement to Tactical Manual F-4B/J Aircraft of 1 Mar 1969
- NAVAIR 01-45HHA-1T Tactical Manual for F-8 Aircraft of 15 Sep 1968
- NAVAIR 01-45HHA-1T Supplement to Tactical Manual for F-8 Aircraft of 1 Sep 1968
- NAVAIR 01-40AV-1T, "A-4/TA-4 Tactical Manual " of 1 Apr 1967
- NAVAIR 01-85ADA-1T, "A-6 Tactical Manual " of 15 Dec 1967
- NAVAIR 01-45AAA-1T, "A-7 Tactical Manual " of 15 Apr 1968
- Weapons System Evaluation Group (WSEG) Secret Report #116, of 20 October 1967,

"Account of F-4 and F-8 Events prior to March 1967 "
- National Security Archive October 29, 2013
- **http://www2.gwu.edu/~nsarchiv/NSAEBB/NSAEBB443/**
- **http://www2.gwu.edu/~nsarchiv/NSAEBB/NSAEBB443/docs/area51_48.PDF**
- http://www2.gwu.edu/-nsarchiv/NSAEBB/NSAEBB443/docs/area51_49.PDF
- **http://www2.gwu.edu/~nsarchiv/NSAEBB/NSAEBB443/docs/area51_50.PDF**
- RED EAGLES, America's Secret MiG by Steve Davies
- America's SECRET MiG Squadron by Gail Peck

Campbell, John M. and Hill, Michael. Roll Call: Thud. Atglen, PA: Schiffer Publishing Ltd., 1996. ISBN 0-7643-0062-8.

Hobson, Chris. Vietnam Air Losses, USAF, USN, USMC, Fixed-Wing Aircraft Losses in Southeast Asia 1961–1973. North Branch, Minnesota: Specialty Press, 2001. ISBN 1-85780-115-6.

- FTD-CR-20-13-69-INT Vol I 3 October 1997
- FTD-CR-20-13-69-INT Vol II August 23, 2000
- FTD-CR-20-02-69 Vol I April 1970 Declassified October 3, 1997
- FTD-CR-20-02-69 Vol II April 1970 Declassified 3 October 1997

The MiG Bandits of Nevada

Red Hats (HAVE DOUGHNUT, HAVE Drill/Ferry, Have Idea) 1968-1993
Robert "Bob" Acosta (1986-1989) – AFSC, Have Idea
Skip Anderson (1973-1976) – AFSC, Have Idea
Robert G. Ashcraft (1968-1969) – TAC, HAVE DOUGHNUT, HAVE Drill/Ferry
John J. Barnoski (1987-1991) – AFSC, Have Idea, 6513TS/CC
Jon S. Beesley (1980-1981) – AFSC, Have Idea
Robert Belt (1989) – TAC, Have Idea
Matt Black (1992-1993) – ACC, Have Idea
James E. "J.B". Brown III (1989-1993) – AFSC/AFMC, Have Idea
David R. Bryant (1983-1986) – USN, Have Idea
Joe Lee Burns (1972-1974) – TAC, Have Idea
Kevin P. Burns (1989-1993) – AFSC/AFMC, Have Idea, 6513TS/CC, 413FLTS/CC
David Carr (1993) – ACC, Have Idea
John H. Casper (1975-1981) – AFSC, Have Idea, 6513TS/CC
Thomas J. Cassidy Jr. (1968) – USN, HAVE DOUGHNUT
Chris Ceplecha (1991-1993) – TAC/ACC, Have Idea
Philip J. Conley Jr. (1978-1982) – AFSC, Have Idea, AFFTC/CC
Fred J. Cuthill (1968-1969) – AFSC, HAVE DOUGHNUT, HAVE Drill/Ferry
Daniel N. Dixon (1992-1993) – USN, Have Idea
Norman K. "Ken" Dyson (1976-1977) – AFSC, Have Idea
John R. "Russ" Easter (1981) – AFSC, Have Idea
David L. Ferguson (1973-1979) – AFSC, Have Idea, 6513TS/CC
Peter B. Field USMC, Have Idea
Mike Ford TAC, Have Idea
Robert S. Frank (1985-1988) – AFSC, Have Idea, 6513TS/CC

Ralph H. Graham (1981-1985) – AFSC, Have Idea, DET3/CC
Steven A. Green (1985-1986) – AFSC, Have Idea
Thomas C. Horne (1993-1995) – AFMC, 413 FLTS/CC
Jerry "Devil" Houston (1969) – USN, HAVE Drill/Ferry
Paul P. Jacobs Jr. AFSC, Have Idea
James A. Jimenez (1991-1993) – AFMC
Maurice B. "Duke" Johnson (1969) – TAC, HAVE Drill/Ferry
Gayland E. Jones AFSC, Have Idea
Joe B. Jordan (1968) – AFSC, HAVE DOUGHNUT
George Kailiwai III AFSC, Have Idea
D. Keator (1980) TAC, Have Idea
John V. Kelley AFSC, Have Idea
William J. "Pete" Knight – AFSC, Have Idea
Fred D. Knox Jr. (1981-1983) – USN, Have Idea
Joseph A. "Broadway Joe" Lanni (1992-1995) – AFMC, Have Idea
Gerald D. Larson (1968) – TAC (1137th SAS) HAVE DOUGHNUT
Rodney K. H. Liu – AFSC, Have Idea
Michael V. Love – AFSC, Have Idea
James D. "Tony" Mahoney (1990-19??) – TAC/ACC, Have Idea
Marvin L. "Roy" Martin (1979-1981) – AFSC, Have Idea
David G. Mazur –AAFSC, Have Idea
Larry D. McClain (1977-1981) – AFSC, Have Idea
Ronald E. "Mugs" McKeown (1969) – USN, HAVE Drill/Ferry
Donald R. McMonagle (1986-1987) – AFSC, Have Idea
Edward T. "Tom" Meschko – AFSC, Have Idea
Robert A. Moseley (1981) – AFSC, Have Idea
George K. Muellner (1979-1982) – AFSC, Have Idea
Thomas A. Morganfeld (1975-1976) – USN, Have Idea
Steven R. Nagel (1977-1979) – AFSC, Have Idea
William E. Nelson AFSC, Have Idea
Dennis L. Nuttbrock AFSC, Have Idea
Craig T. Otto AFSC, Have Idea
Philip C. Pirozzi (1991-1992) – USN, Have Idea
Michael C. "Bat" Press (1972-1977) – TAC, Have Idea
Joe M. Roberts AFSC, Have Idea
Dennis F. Sager (1991-1993) – AFSC/AFMC, Have Idea
Paul W. Savage (1986) – AFSC, Have Idea
Wendell H. Shawler (1969-1970) – AFSC, HAVE Drill/Ferry, Have Privilege
David L. "DL" Smith (1972-1975) – TAC, Have Idea
Thomas P. Stafford (1975-1978) – AFSC, Have Idea, AFFTC/CC
Norman L. Suits (1971-1973) – AFSC, Have Idea
Thomas S. Swalm (1969) – TAC, HAVE Drill/Ferry
Dane C. Swanson (1985-1989) – USN, Have Idea
Paul D. Tackabury (1980-1981) – AFSC, Have Idea
John A. "Ashby" Taylor (1981-1985) – AFSC, Have Idea, 6513TS/CC
Foster S. "Tooter" Teague (1969) – USN, HAVE Drill/Ferry

James W. Tilley II (1980-1984) – AFSC, 6513TS/CC
William T. "Ted" Twinting (1968) – AFSC, HAVE DOUGHNUT
Teddy Varwig (1991-1992) – TAC, Have Idea
Ken Wallace (1988) – USN, Have Idea
Charles P. "Pete" Winters (1971-1983) – AFSC, Have Idea, DET3/CC
Otto J. Waniczek (1979-1982) – AFSC, Have Idea, FTE
Mike Welch (1969) – USN HAVE Drill/Ferry
James H. Wisneski (1981-1986) – AFSC, Have Idea
Red Eagles (Constant Peg, Have Idea) 1977-1988
David F. "Blazo" Bland (1983-1985) – TAC, Constant Peg (Bandit 32)
Thomas V. "Boomer" Boma (1986-19??) – TAC, Constant Peg (Bandit 59)
Melvin Hugh "Bandit" Brown (1979) – USN, Constant Peg (Bandit 12)
Steven R. "Brownie" Brown (1983-1986) – TAC, Constant Peg (Bandit 33)
Guy A. Brubaker (1984-1986) – USN, Constant Peg (Bandit 48)
Leonard J. Bucko (1981-1983) – USMC, Constant Peg (Bandit 22)
Herbert J. "Hawk" Carlisle (1986-1988) – TAC, Constant Peg (Bandit 54)
Ricardo M. "Rick" Cazessus (1986-19??) – TAC, Constant Peg (Bandit 63)
Charles T. "Chuck" Corder (1979-1983) – TAC, Constant Peg (Bandit 15)
Evan M. Chanik Jr. (1985-1988) – USN, Constant Peg (Bandit 52)
Robert E. Craig (1984-1987) – TAC, Constant Peg (Bandit 43)
Robert E. "Sundance" Davis (1987-198?) – USN, Constant Peg (Bandit 64)
Thomas E. "Gabby" Drake (1984-1987) – TAC, Constant Peg (Bandit 42)
James B. "Bear" Evans (1987-198?) – TAC, Constant Peg (Bandit 67)
Glenn F. "Pappy" Frick (1972-1978) – TAC, Have Idea, Constant Peg, 4477TEF/CC (Bandit 4)
Nickie J. Fuerst (1987-198?) – TAC, Constant Peg (Bandit 65)
Francis K. "Paco" Geisler (1983-1986) – TAC, Constant Peg (Bandit 35)
George S. "G2" Gennin (1982-1984) – TAC, Constant Peg, 4477TES/CC (Bandit 31)
Thomas A. Gibbs (1980-1982) – TAC, Constant Peg, 4477TES/CC (Bandit 21)
James W. "Wiley" Green (1982-1984) – TAC, Constant Peg (Bandit 26)
Charles J. "Heater" Heatley III (1977-1981) – USN, Constant Peg (Bandit 8)
Earl J. Henderson (1979-1980) – USN, Have Idea, Constant Peg 4477TEF/CC, 4477TES/CC (Bandit 14)
Gerrald D. "Huffer" Huff (1977-1980) – TAC, Constant Peg (Bandit 6)
Ronald W. Iverson (1975-1979) – TAC, Have Idea (Bandit 2)
Timothy R. "Stretch" Kinney (1984-1987) – TAC, Constant Peg (Bandit 45)
Dudley A. Larsen (1986-19??) – TAC, Constant Peg (Bandit 60)
Selvyn S. "Sel" Laughter (1980-1982) – USN, Constant Peg (Bandit 18)
Martin S. Macy (1985-1988) – USMC, Constant Peg (Bandit 49)
James D. "Tony" Mahoney (1987-1990) – USN, Constant Peg (Bandit 62)
John T. "Jack" Manclark (1985-1987) – TAC, Constant Peg, 4477TES/CC (Bandit 51)
John C. "Flash" Mann (1986-1988) – TAC, Constant Peg (Bandit 56)
James D. "Thug" Matheny (1982-1985) – TAC, Constant Peg (Bandit 27)
Robert "Kobe" Mayo (1972-1976) – TAC, Have Idea (Bandit 1)
David J. "Marshall" McCloud (1979-1981) – TAC, Constant Peg (Bandit 6)
Daniel R. "Bad Bob" McCort (1984-1986) – USN, Constant Peg (Bandit 44)

Douglas M. Melson (1987-1988) – TAC, Constant Peg (Bandit 68)
Thomas A. Morganfeld (1977-1980) – USN, Have Idea, Constant Peg (Bandit 7)
Brian E. McCoy (1986-19??) – TAC, Constant Peg (Bandit 53)
Alvin D. "Devil" Muller (1974-1978) – TAC, Have Idea (Bandit 3)
Clem B. "Buffalo" Myers (1980-19??) – TAC, Constant Peg (Bandit 19)
John B. "Black" Nathman (1982-1984) – USN, Constant Peg (Bandit 29)
Joseph L. "Jose" Oberle (1978-1982) – TAC, Constant Peg (Bandit 5)
Stanley R. "Swish" O'Connor (1988-1990) – USN, Constant Peg (Bandit1969)
Gaillard R. "Evil" Peck (1978 – 1979) – TAC, Constant Peg, 4477TEF/CC (Bandit 9)
Edward D. "Hog" Phelan (1984-1988) – TAC, Constant Peg (Bandit 47)
Mark F. "Toast" Postai (1981-1982) – TAC, Constant Peg (Bandit 25)
Michael C. "Bat" Press (1972-1977, 1980-1981) – TAC, Have Idea, Constant Peg (Bandit 20)
Orville Prins (1982-1985) – TAC, Constant Peg (Bandit 30)
James A. "Rookie" Robb (1983-1985) – USN, Constant Peg (Bandit 38)
Shelley S. "Scotty" Rogers (1986-19??) – TAC, Constant Peg (Bandit 55)
Michael C. Roy (1983-1985) – TAC, Constant Peg (Bandit 36)
John B. Saxman (1983-1986) – TAC, Constant Peg (Bandit 34)
Michael R. "Scotty" Scott (1979-1983, 1987-1990) – TAC, Constant Peg, 4477TES/CC (Bandit 14)
Keith E. Shean (1980-1982) – TAC, Constant Peg (Bandit 17)
Robert "Catfish" Sheffield (1979-1982) – TAC, Constant Peg (Bandit 16)
Larry T. "Shy" Shervanick (1982-1985) – TAC, Constant Peg (Bandit 28)
Cary A. "Dollar" Silvers (1986) – USN, Constant Peg (Bandit 61)
Michael G. Simmons (1986-19??) – TAC, Constant Peg (Bandit 57)
John "Grunt" Skidmore (1984-1985) – TAC, Constant Peg (Bandit 41)
Paul R. "Stook" Stucky (1983-1986) – TAC, Constant Peg (Bandit 40)
Charles E. "Smokey" Sundell (1986-19??) – TAC, Constant Peg (Bandit 58)
Russell M. "Bud" Taylor (1981-1983) – USN, Constant Peg (Bandit 23)
Sam C. Therrien (1987-198?) – TAC, Constant Peg (Bandit 66)
Frederick H. "T-Bear" Thompson (1985-198?) – TAC, Constant Peg (Bandit 50)
George C. "Cajun" Tullos (1983-1985) – USMC, Constant Peg (Bandit 37)
James M. "Monroe" Watley (1981-1984) – TAC, Constant Peg (Bandit 24)
Philip W. "Hound Dog" White (1984-1986) – TAC, Constant Peg, 4477TES/CC (Bandit 46)
Karl F. "Harpo" Whittenberg (1979-19??) – TAC, Constant Peg (Bandit 11)
Robert J. "Z-Man" Zettel (1983-1986) – TAC, Constant Peg (Bandit 39)

US planes lost in Vietnam

United States Air Force

All told, the US Air Force flew 5.25 million sorties over South Vietnam, North Vietnam, northern and southern Laos, and Cambodia, losing 2,251 aircraft: 1,737 to hostile action and 514 in accidents, 110 of the losses were helicopters and the rest fixed-wing. A ratio of roughly 0.4 losses per 1,000 sorties compared favorably with a 2.0 rate in Korea and the 9.7 figure during World War II.

Sources for USAF figures:

USAF Operations Report, 30 November 1973

Campbell, John M. and Hill, Michael. Roll Call: Thud. Atglen, PA: Schiffer Publishing Ltd., 1996. ISBN 0-7643-0062-8.

Hobson, Chris. Vietnam Air Losses, USAF, USN, USMC, Fixed-Wing Aircraft Losses in Southeast Asia 1961–1973. North Branch, Minnesota: Specialty Press, 2001. ISBN 1-85780-115-6.

USAF fixed-wing

Downed USAF Douglas A-1E, pilot later awarded the Medal of Honor

A-1 Skyraider-- --191 total, 150 in combat

First loss A-1E 52-132465 (1st Air Commando Squadron, 34th TG) shot down during the night of 28–29 August 1964 near Bien Hoa, SVN

Final loss A-1H 52-139738 (1st Special Operations Squadron, 56th Special Operations Wing) which was shot down 28 September 1972 (pilot rescued by an Air America helicopter).

A-7D Corsair II-- --6 total, 4 combat

–First loss 71–0310 (355th Tactical Fighter Squadron, 354th TFW) on 2 December 1972 shot down on a CSAR mission in Laos (Capt Anthony Shine KIA).

–71-0312 (353d TFS) mid-air collision with an FAC O-1 Bird Dog in Laos on 24 December 1972, (Capt Charles Riess PoW)

-71-0316 (355th TFS) operational loss (non-combat) crash in Thailand on 11 January 1973 (Pilot Rescued)

-70-0949 (354th TFW) shot down Laos on 17 February 1973 (Maj. J J Gallagher Rescued)

-71-0305 (3rd TFS, 388th TFW) shot down in Cambodia on 4 May 1973 (1Lt T L Dickens Rescued)

Final loss 70–0945 (354th TFW) shot down in Cambodia on 25 May 1973 (Capt Jeremiah Costello KIA)

A-26 Invader-- --22 total

–First loss B-26B 44-35530 (Detachment 2A, 1st ACG) shot down in IV CTZ on the night of 4–5 November 1962 killing the 3 crew.

–Final loss A-26A 64-17646 (609th SOS, 56th SOW) lost over Laos on the night of 7–8 July 1969 killing both crewmen.

A-37 Dragonfly-- --22 total

–First loss 1967; final loss 1972

Wing of downed USAF warplane

AC-47 Spooky-- --19 total, 12 in combat

–First loss 1965, final loss 1969

AC-119 Shadow/Stinger-- --6 total, 2 in combat

–First loss AC-119G 52-5907 (Det.1, 17th SOS, 14th SOW) which crashed on take-off from Tan Son Nhut, SVN on 11 October 1969 killing 6 of the 10 crewmen.

–Final loss 1971

AC-130 Spectre-- --6 total, all combat.

–First loss AC-130A 54-1629 (16th SOS, 8th TFW) hit by 37 mm AAA over Laos and crash-landed at Ubon RTAFB, 2 crewmen died (one died of injuries before reaching Ubon) but 11 others survived.

–Final loss 1972

B-52 Stratofortress-- --31 total, 17 in combat

-First losses were operational (non-combat) mid-air collision 2 B-52F 57-0047 and 57-0179 (441st Bomb Squadron, 320th Bomb Wing), 18 June 1965, South China Sea during air refueling orbit, 8 of 12 crewmen killed

-Final loss B-52D 55-0056 (307th Bomb Wing Provisional) to SAM 4 January 1973, crew rescued from Gulf of Tonkin

B-57 Canberra-- --56 total, 38 in combat

-First loss 1964, final loss 1970

C-5A Galaxy-- --1 total, 0 in combat. Crashed while attempting emergency landing at Tan Son Nhut AB 4 April 1975, as part of Operation Babylift. Five of the eight US Military women killed during the Vietnam War, were aboard this airplane.

C-7 Caribou-- --19 total, 9 in combat

-First lost C-7B 62-4161 (459th Tactical Airlift Squadron, 483d Tactical Airlift Wing) which was hit by a US 155 mm shell on 3 August 1967 in SVN killing the 3 crewmen. Note: there were two fatal crashes during Operation Red Leaf transition training of USAF crews in Army CV-2's, on 4 and 28 October 1966.

-Final loss was C-7B 62-12584 (483d TAW) which crashed in SVN on 13 January 1971, all 4 crewmen survived.

C-47 Skytrain-- --21 total

-A C-47 was very first USAF aircraft lost in the SEA conflict, C-47B 44-76330 (315th Air Division) on TDY at Vientiane, Laos which was shot down by the Pathet Lao on 23 March 1961 killing 7 of the 8 crewmen. The sole survivor, US Army Maj. Lawrence Bailey was captured and held until August 1962.

-Final loss EC-47Q 43-48636 (361st Tactical Electronic Warfare Squadron, 56th SOW) shot down in Laos on the night of 04/5 4–5 February 1973 killing all eight crewmen.

C-123 Provider-- --53 total, 21 in combat

-First loss was C-123B 56-4370 attached to the 464th TAW which came down on an Operation Ranch Hand (defoliation) training flight between Bien Hoa and Vung Tau, SVN on 2 February 1962

-Final loss 1971

C-130 Hercules-- --55 total, 34 in combat

-First loss was C-130A 57-0475 (817th Troop Carrier Squadron, 6315th Operations Group) on 24 April 1965, a Blind Bat flareship that crashed into high ground near Korat Royal Thai Air Force Base, Thailand, attempting to land in bad weather with a heavy load, two engine failures, and low fuel, killing all six crewmen. This was the 14th recorded loss of a C-130 to all causes.

-Final loss C-130E 72-1297 (314th TAW) destroyed by rocket fire at Tan Son Nhut AB on 28 April 1975.

C-141 Starlifter-- --2 total, 0 combat

-C-141A 65-9407 (62d Military Airlift Wing) destroyed in a night runway collision with a USMC A-6 at Danang, SVN on 23 March 1967 killing 5 of the 6 crewmen.

-C-141A 66-0127 (4th Military Airlift Squadron, 62d MAW) crashed soon after take-off from Cam Ranh Bay, SVN on 13 April 1967 killing 6 of the 8 man crew.

E/RB-66 Destroyer—14 total

-First loss was RB-66B 53-0452 (Det 1, 41st Tactical Reconnaissance Squadron, 6250th Combat Support Group) which crashed 22–23 October 1965 west of Pleiku, SVN killing the crew.

-Final loss EB-66B 42nd TEWS, 388th TFS lost to engine failure on 23 December 1972 during Operation Linebacker II. 3 crewmen were KIA.

EC-121 BatCat-- --2 total, 0 combat

-EC-121R 67-24193 (554th Reconnaissance Squadron, 553d RW) crashed 25 April 1969 on take-off in a thunderstorm from Korat RTAFB, killing all 18 crewmen.

-EC-121R 67-21495 (554th RS) crashed on approach to Korat RTAFB on 6 September 1969, 4 of the 16 men killed.

F-4 Phantom II-- --445 total, 382 in combat

-First loss was operational (non-combat), F-4C 64-0674 (45TH TFS, 15th TFW) which ran out of fuel after strike in SVN on 9 June 1965; first combat loss F-4C 64-0685 (45th TFS, 15th TFW) shot down Ta Chan, NW NVN on 20 June 1965. 9 of the losses were parked aircraft struck by rockets.

-Final loss, F-4D 66-8747 (432d TRW) on 29 June 1973.

F-5 Freedom Fighter-- --9 total

-First loss 1965, final loss 1967

F-100 Super Sabre-- --243 total, 198 in combat

-First loss 1964, final loss 1971

F-102 Delta Dagger-- --14 total, 7 combat

-First loss 1964, final loss 1967. 4 of the combat losses were parked aircraft

F-104 Starfighter-- --14 total, 9 combat

-First loss 1965, final loss 1969

F-105D Thunderchief-- --335 total, 283 in combat

-First loss 62–4371 (36th TFS, 6441st TFW) written off from battle damage over Laos 14 August 1964, at Korat, Thailand

-Final loss 61–0153 (44th TFS, 355th TFW) shot down Laos 23 September 1970, pilot Capt. J. W. Newhouse rescued

F-105F/G Thunderchief: "Wild Weasel," "Ryan's Raiders," "Combat Martin"-- --47 total, 37 combat

-First loss EF-105F 63-8286 (13th TFS, 388th TFW) shot down by AAA RP-6 July 1966, Maj. Roosevelt Hestle, and Capt. Charles Morgan KIA

-Last loss F-105G 63-8359 (Det.1 561st TFS, 388th TFW) shot down by SAM 16 November 1972, RP-3, crew rescued

F-111A "Aardvark"-- --11 total, 6 in combat

-First loss mission-related TFR failure, 66-0022 (428th TFS 474th TFW, Project Combat Lancer), 28 March 1968, Maj. H.E. Mccann and Capt. D.L. Graham MIA

-Final loss 67–0111 (474th TFW) mid-air collision over Cambodia, 16 June 1973, both crewmen rescued

HU-16 Albatross-- --4 total, 2 combat

-51-5287 to unknown cause 19 June 1965

-51-0058 to unknown cause 3 July 1965

-51-0071 (33d ARRS) shot down by AAA 14 March 1966, two crewmen killed

-51-7145 (37th ARRS) disappeared 18 October 1966, 7 crewmen KIA-BNR

KB-50 Superfortress tanker-- --1 total, 0 combat

-Only loss KB-50J 48-0065 (421st Air Refueling Squadron Detachment) at Takhli RTAFB, which crashed in Thailand on 14 October 1964, all 6 crewmen, survived.

KC-135 Stratotanker-- --3 total, 0 combat

-Two crashes in 1968, one 1969, all operational (non-combat)

O-1 Bird Dog-- --172 total, 122 in combat

-First loss 1963, final loss 1972

O-2 Skymaster-- --104 total, 82 in combat

-First loss 1967, final loss 1972

OV-10 Bronco-- --63 total, 47 in combat

First loss 1968, final loss 1973

QU-22 Pave Eagle-- --8 lost, 7 in combat

-First loss YQU-22A 68-10531 (554th RS, 553d RW) crashed due to engine failure on 11 June 1969

-Final loss QU-22B 70-1546 (554th RS) on 25 August 1972, pilot killed.

RF-4C Phantom II-- --83 total, 76 in combat

-First loss 1966, final loss 1972

RF-101 Voodoo-- --39 total, 33 in combat

-First loss 1964, final loss 1968

SR-71A Blackbird-- --2 total, 0 combat

-64-17969 (Det OL-8, 9th Strategic Reconnaissance Wing) suffered engine failure over Thailand on 10 May 1970, both crewmen ejected safely

-64-17978 (Det OL-KA, 9th SRW) crashed on landing at Kadena, Okinawa on 20 July 1972, both crewmen survived

T-28 Trojan-- --23 total

-First loss 1962, final loss 1968

U-2C "Dragon Lady"-- --1 total, 0 combat

-Only loss 56–6690 (349th Strategic Reconnaissance Squadron 100th SRW) which crashed on 8 October 1966 near Bien Hoa, SVN, Maj. Leo J Stewart ejected and rescued.

U-3B Blue Canoe-- --1 total, 1 combat

-Only loss 60–6058, destroyed on the ground during a VC attack on Tan Son Nhut, SVN on 14 June 1968.

U-6A Beaver   1 total, 0 combat

Only loss 51-15565 (432d Tactical Reconnaissance Wing) which crashed in Thailand 28 December 1966, both crewmen survived.

U-10D Courier-- --1 total, 1 combat

-63-13102 (5th SOS, 14th SOW) shot down 14 August 1969 near Bien Hoa killing 1/Lt Roger Brown.

CH/HH-3 Jolly Green Giant-- --34 total, 25 in combat

-First loss CH-3E 63-9685 (38th ARRS) to AAA North Vietnam on 6 November 1965, three crewmen POW, one rescued

-Last loss HH-3E 65-12785 (37th ARRS) 21 November 1970, combat-assaulted inside Son Tay POW camp (Operation Ivory Coast) and deliberately destroyed by US Special Forces

HH-43B Pedro-- --13 lost, 8 in combat

-First loss 63–9713 (38th ARRS) damaged by fire 2 June 1965, crew rescued and aircraft destroyed to prevent its capture

-Final loss 60–0282 (38th ARRS) crashed Cam Ranh Bay 7 August 1969, crew rescued

CH/HH-53 Super Jolly-- --27 total, 17 in combat

-First loss HH-53C 66-14430 (40th ARRS) in Laos, damaged by gunfire 18 January 1969 crew rescued and aircraft destroyed by bombing to prevent capture

-Last losses four CH-53's (68-10925, −10926, −10927, 70–1627 all from 21st SOS, 56th SOW) to AAA on 15 May 1975, Koh Tang, Kampuchea, (Mayaguez incident final aircraft losses of Vietnam War)

UH-1 Iroquois-- --36 total

United States Navy

Twenty-one aircraft carriers conducted 86 war cruises and operated 9,178 total days on the line in the Gulf of Tonkin. 530 aircraft were lost in combat and 329 more to operational causes. Resulting in the deaths of 377 naval aviators, with 64 airmen reported missing and 179 taken prisoner-of-war.

Sources for USN carrier-based figures:

Francillon, René. Tonkin Gulf Yacht Club: US Carrier Operations off Vietnam, Naval Institute Press (1988) ISBN 0-87021-696-1

USN fixed-wing carrier-based

A-1 Skyraider—65 total, 48 in combat

-First loss A-1H 139760 (VA-145, USS Constellation), to AAA 5 August 1964, Lt.j.g. R. C. Sather KIA (Body recovered in 1985)

-Final loss A-1H 134499 (VA-25, USS Coral Sea), to MiG 14 February 1968, Lt.j.g. J. P. Dunn MIA

A-3 Skywarrior—7 total, 2 in combat

-First loss A-3B 142250 (VAH-4, USS Hancock), operational loss (non-combat) 22 December 1964, 3 rescued, 1 killed

-Final loss A-3B 144627 (VAH-4, USS Kitty Hawk), AAA 8 March 1967, 3 crewmen KIA

A-4 Skyhawk—282 total, 195 in combat

-First loss A-4C 149578 (VA-144, USS Constellation), AAA 5 August 1964, Lt.j.g. Everett Alvarez POW (second longest held prisoner)

-Final loss A-4F 155021 (VA-212, USS Hancock), AAA 6 September 1972, pilot rescued

A-6 Intruder—62 total, 51 in combat

-First loss A-6A 151584 (VA-75, USS Independence), own bomb detonation Laos 14 July 1965, crew rescued

-Final loss A-6A 157007 (VA-35, USS America), AAA South Vietnam 24 January 1973, crew rescued

A-7 Corsair—100 total, 55 in combat

-First loss A-7A 153239 (VA-147, USS Ranger), SAM North Vietnam 22 December 1967, LCDR J.M. Hickerson POW

-Final loss A-7E 156837 (VA-147, USS Constellation), operational loss (non-combat) 29 January 1973, pilot missing

C-1 Trader—4 total, 0 in combat

-C-1A 146047 (VR-21, USS Independence), non-combat 30 August 1965, 7 passengers and crew rescued

-C-1A 136784 (VR-21, USS Independence), operational loss (non-combat) 12 September 1965, 9 passengers and crew rescued, 1 killed

-C-1A 146016 (Composite Squadron Five VC-5), operational loss (non-combat) 8 August 1967, 3 passengers and 2 crew rescued

-C-1A 146054 (Carrier Air Wing 11, Kitty Hawk), operational loss (non-combat) 16

January 1968, 7 passengers and crew rescued, 3 killed

C-2 Greyhound—1 total, 0 in combat
-Sole loss C-2A 155120 (VRC-50, USS Ranger), Gulf of Tonkin crash 15 December 1970, 9 passengers and crew killed

E-1 Tracer—3 total, 0 in combat
-First loss E-1B 148918 (VAW-12, USS Independence), operational loss (non-combat) 22 September 1965, crew rescued
-Final loss E-1B 148132 (VAW-111, USS Oriskany), operational loss (non-combat) 8 October 1967, 5 crewmen killed

E-2 Hawkeye—2 total, 0 in combat
-E-2A 151711 (VAW-116, USS Coral Sea), 8 April 1970, 5 crewmen killed
-E-2B 151719 (VAW-115, USS Midway), 11 June 1971, 5 crewmen missing

EKA-3 Skywarrior-- --2 lost, 0 in combat
-EKA-3B 142400 (VAQ-132, USS America), operational loss (non-combat) 4 July 1970, 3 rescued
-EKA-3B 142634 (VAQ-130, USS Ranger), operational loss (non-combat) 21 January 1973, 3 crewmen killed

EA-1 Skyraider—4 total, 1 in combat
-First loss EA-1E 139603 (VAW-111, USS Yorktown), operational loss (non-combat) 15 April 1965, crew rescued
-Final loss EA-1F 132543 (VAW-13, USS Franklin D Roosevelt), operational loss (non-combat) 10 September 1966, crew rescued

F-4 Phantom—138 total, 75 in combat
-First loss F-4B 151412 (VA-142, USS Constellation), operational loss (non-combat) 13 November 1964, crew rescued
-Last combat loss (also last USN combat loss of war) F-4J 155768 (VF-143, USS Enterprise), AAA South Vietnam 27 January 1973, CDR H.H. Hall and LCDR P.A. Keintzer POW
-Final loss F-4J 158361 (VF-21, USS Ranger), operational loss (non-combat) 29 January 1973, crew killed

F-8 Crusader—118 total, 57 in combat
-First loss F-8D (VF-111, USS Kitty Hawk), to AAA over Laos 7 June 1964, LCDR C.D. Lynn rescued
-Final loss (operational) F-8J 150887 (VF-191, USS Oriskany) 26 November 1972, pilot rescued

KA-3 Skywarrior- --2 lost, 0 in combat
-KA-3B 142658 (VAH-4, USS Oriskany), operational loss (non-combat) 28 July 1967, 1 crewmen rescued, 2 killed
-KA-3B 138943 (VAH-10, USS Coral Sea), operational loss (non-combat) 17 February 1969, 3 crewmen killed

RA-5 Vigilante—27 total, 18 in combat
-First loss RA-5C 149306 (RVAH-5, USS Ranger), operational loss (non-combat) 9 December 1965, 2 crewmen killed
-Final loss RA-5C 156633 (RVAH-13, USS Enterprise), to MiG-21 North Vietnam 28 December 1972, LCDR A.H. Agnew POW, Lt. M.F. Haifley KIA

RF-8 Crusader—29 total, 19 in combat

-First loss RF-8A (Det. C VFP-63, USS Kitty Hawk), 6 June 1964, to AAA in Laos, Lt. C. F. Klusmann POW

-Final loss RF-8G 144608 (VFP-63, USS Oriskany), operational loss (non-combat) 13 December 1972, pilot rescued

S-2 Tracker—5 total, 3 in combat

-First loss S-2D 149252 (VS-35, USS Hornet), unknown combat loss 21 January 1966, 4 crewmen MIA

-S-2E 152351 (VS-21, USS Kearsarge), combat loss 11 October 1966, 4 crewmen KIA

-US-2C 133365 (VC-5, NAS Atsugi, Japan), combat loss 13 May 1967, 2 crewmen KIA

-US-2C 133371 (VC-5, USS Hornet), operational loss (non-combat) 27 September 1967, crew rescued

-Final loss S-2E (VS-23, USS Yorktown), unknown combat loss 17 March 1968, 4 crewmen KIA

USN fixed-wing shore-based

C-47 Skytrain (1)

OV-10 Bronco (7)

P-2 Neptune (4)

P-3 Orion (2)

USN rotary-wing

SH-2/UH-2 Sea Sprite-- --12 lost, 0 in combat

-First loss UH-2A 149751 (HC-1, USS Hancock), operational loss (non-combat) 10 January 1966, 4 crewmen rescued

-Final loss UH-2C 149767 (HC-1, USS Bon Homme Richard), operational loss (non-combat) 10 August 1969, 4 crewmen rescued

SH-3 Sea King-- --20 lost, 8 in combat

-First loss SH-3A 148993 (HS-2, USS Hornet), AAA North Vietnam 13 November 1965, 4 crewmen rescued

-Final loss SH-3D 156494 (HS-7, USS Saratoga), operational loss (non-combat) 31 December 1972, crew rescued

United States Marine Corps

US Marine Corps aircraft lost in combat included 193 fixed-wing and 270 rotary wing aircraft.

USMC fixed-wing

A-4 Skyhawk—81 lost

A-6 Intruder—25 lost

C-117 Skytrain—2 lost

EA-6A Intruder—2 lost

EF-10 Skynight—5 lost

F-4 Phantom—95 lost, 72 combat

F-8 Crusader—21 lost

KC-130 Hercules—4 lost

O-1 Bird Dog—7 lost

OV-10 Bronco—10 lost

RF-4 Phantom—4 lost

RF-8 Crusader—1 lost

TA-4 Skyhawk—10 lost
TF-9 Cougar—1 lost
USMC rotary-wing

Downed UH-1 Iroquois
AH-1 Cobra – 7
HUS-1 – 75
UH-1E Huey –1969
CH-37 Mojave – 1
CH-46D Sea Knight – 109
CH-53 Sea Stallion – 9

United States Army
USA fixed-wing
OV-1A Mohawk – 3 lost
OV-1B Mohawk – 2 lost
O-1 Bird Dog – 297 lost

More than 67 OV-1afterburner/C/D series aircraft were lost in combat, many of them in Laos and North Vietnam. In just seven months in 1966, the 20th ASTA/131 Aviation Company (Mohawks) lost 6 OV-1As, 20 OV-1Bs, and 2 OV-1C.

USA rotary-wing
5,086 (which include not in addition to the above statistics)

1 Bell 205 (Air America)
270 AH-1G
1 BELL
14 CH-21C
2 CH-34
1 CH-37B
1 CH-37C
83 CH-47A destroyed
20 CH-47B destroyed
29 CH-47C destroyed
9 CH-54A destroyed
3 H-13D destroyed
2 H-37A destroyed
147 OH-13S destroyed
93 OH-23G destroyed
45 OH-58A destroyed
842 OH-6A destroyed
60 UH-1 destroyed
357 UH-1B destroyed
365 UH-1C destroyed
886 UH-1D destroyed
90 UH-1E destroyed
18 UH-1F destroyed

1313 UH-1H destroyed
176 UH-34D destroyed
0

About the Author

Thornton D. "TD" Barnes, born in 1937m grew up on the family's homestead ranch at Dalhart, Texas. He joined the Oklahoma National Guard before graduating from High School at Mountain View, Oklahoma. After graduation, he immediately joined the Army where he served in Intelligence in Korea. Thereafter, Barnes attended numerous electronics schools in Army air defense missiles and radar at Fort Bliss, Texas. He served in Germany and then attended OCS at Fort Sill, Oklahoma where he sustained career-ending injuries during survival training. He participated as a field engineer on the NASA High Range in Nevada during the test flights of the X-15, XB-70, Lifting Bodies, and Lunar landing Vehicles. He participated in the NERVA project at Jackass Flats in Nevada developing a nuclear rocket engine for manned flight to Mars. From there, the CIA recruited him for special projects at Area 51 in Nevada where he participated in CIA Project Oxcart developing the A-12 Blackbird surveillance planes. At Area 51, Barnes exploited and reverse engineered Russian MiG aircraft and radar that became the genesis of the Red Eagle MiG Aggressor Squadron of Constant Peg at the National Tonopah Testing Range in Nevada, Navy's Top Gun Fighter Weapons School, and the USAF Red Flag exercises that continue today. He also participated in the early development of stealth technology at Area 51.

Barnes left government service to form an oil and gas exploration company. His success in business expanded into the mining of uranium and gold. Since retirement, Barnes works with the CIA, various universities, and the Library of Congress to record the history of the classified projects in which he worked. Barnes is president of Roadrunners Internationale, a social organization of the men who worked for the CIA at Area 51. He served as a cofounder and the first director of the Nevada Aerospace Hall of Fame where he focused on educating the youth of Nevada and honoring the legends of aerospace and aviation in that state.

In 2007, the CIA brought Barnes back to CIA during declassification of some of the projects in which he worked at Area 51. In 2010, the CIA declassified Barnes' personal participation in these projects at Area 51. Since then, the documentary "Area 51 Declassified" by the National Geographic Channel extensively identified Barnes a top player in the accomplishments of Area 51. Numerous documentaries by the History Channel, the Discovery Channel, the Travel Channel, and others have also filmed him. The best-selling book by aviation writer Annie Jacobsen focused around the declassified history of Barnes' activities at Area 51.

Barnes is the author of two nonfiction books based upon the recently declassified activities occurring in Nevada. "CIA Bride" is a biography about the CIA commander at Area 51 during Project OXCART and his wife who also worked for the CIA. "My Odyssey to Area 51" is Barnes' memoirs taking him from his family's ranch in New Mexico to the CIA's Paradise Ranch at Area 51. He recently published the first two books of a series titled "EmP." EmP - Nuclear Winter about survival following a nuclear war based upon his experiences at the Atomic Proving Grounds in Nevada, and EmP - Nuclear Spring that picks up the story four years later. During the spring of 2014, Barnes expects to release Book 3 of the series and a nonfiction account of his participation in the Soviet MiG exploitation projects at Area 51. Barnes lives in Henderson, Nevada.

Connect with the Author Online:
Amazon: http://www.amazon.com/-/e/B0086JXXTC
Facebook: **https://www.facebook.com/CodenameThunder?v=info**
Blog: **http://td-barnes.com/blog/?p=231**
Website: http://td-barnes.com/
Linked-In: **http://www.linkedin.com/profile/edit?trk=tab_pro**

Other Books by Author:

http://www.amazon.com/-/e/B0086JXXTC

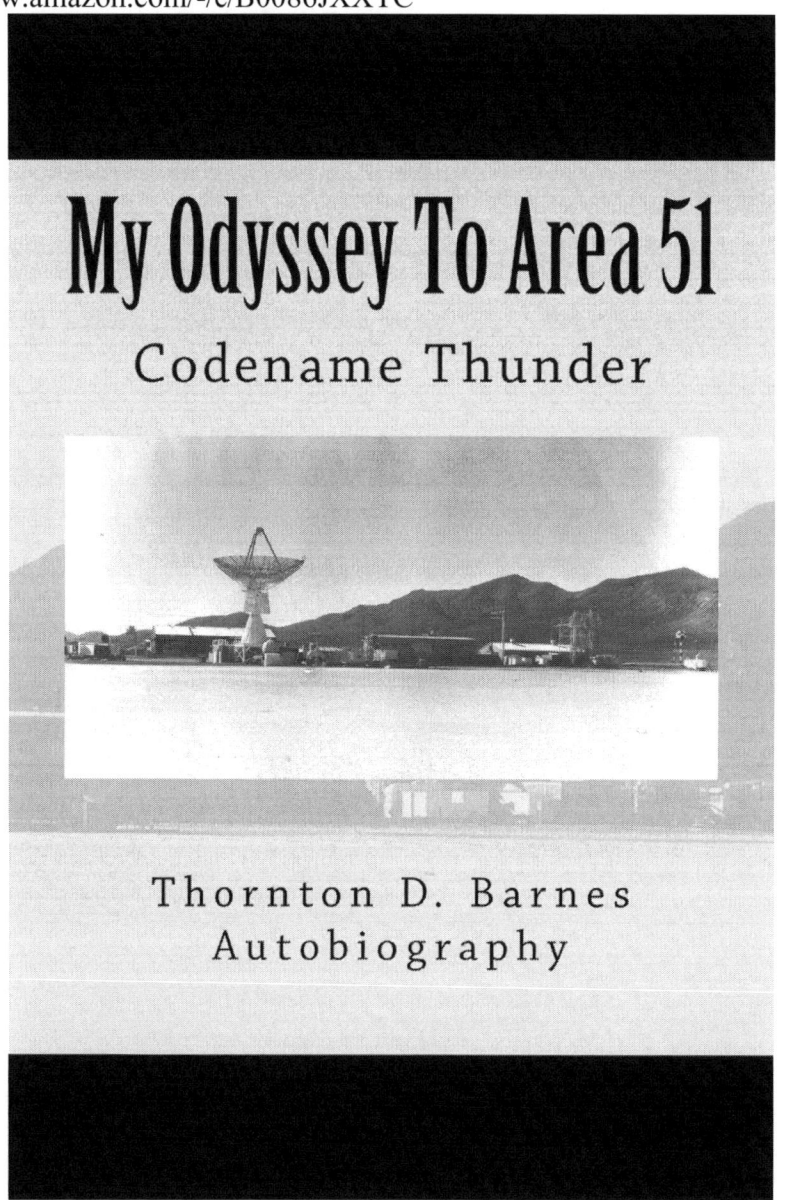

My Odyssey to Area 51s
While working for the CIA at Area 51 Barnes used the code-name "Thunder." My Odyssey to Area 51 follows his experiences on the NASA High Range and Groom Lake where others and he pioneered the NASA X-15, XB-70, and Lifting Bodies. My Odyssey to Area 51 covers his recruitment for the CIA A-12 Oxcart at Groom Lake where he remained as Special Projects cadre for the Soviet MiG exploitation programs, the early days of stealth. The named icons of aviation and business are the eagles with which he soared. The content is declassified.

EMP - Nuclear Winter

Book 1 of EMP Series

Thornton D. Barnes

EMP - Nuclear Winter is Book One of the EMP series. It is a work of fiction because to date the world has not experienced an EMP, electromagnetic pulse attack or a nuclear winter. However, it is merely a matter of time before a group or nation of radicals will use such a dreadful device in a terrorist attack on a major American city with the same effects described in this book.

The scenario for this EMP story models after what is happening now and such an attack could occur at any time. Selecting the state of Nevada the venue makes the story more realistic by affiliating the characters with actual technological activities occurring at the Nevada National Security Site and at the Area 51 technological laboratories. EMP takes the reader through the attack and realistically addresses logistics, security, survivor selection, social, cultural, education, and other issues required to prepare a society capable of rebuilding a nation devoid of the technology they once depended on.

When the reader finishes reading EMP, he or she will wonder why our advanced technological world of social networking is not talking about what will occur when the terrorist organization Al Qaeda, a radical religious group, or rogue nations North Korea or Iran hits the United States with an EMP attack. Even more frightening is no one is preparing the surviving citizens for the day when a radical terrorist cell delivers and detonates an EMP device on a freighter arriving in a major city. This possibly occurring is not a matter of if; it is a matter of when. No one talks about the consequences of such an attack because they are so horrific. No one is preparing defenses for this occurrence, yet the costs of such a defense pales when compared to the long-lasting devastation to the human race because of such an attack. They should do something to protect the people, but are not doing so.

Warning - you will find in this book that the aftermath of an EMP attack is not pretty.

EMP - Nuclear Spring

Book 2 of EMP Series

TD Barnes

    EMP - Nuclear Spring is Book Two of the EMP series. Four years have passed since the EMP attack and the nuclear bombs that followed. For the first time, the survivors are able to come out of the Yucca Mountain underground complex initially constructed to store nuclear waste. They find themselves under surveillance by Islamic activists of the Islamic Brotherhood who survived in Central America. The Brotherhood is moving into Nevada to take control of the Hoover Dam, which gives the Brotherhood control of the Colorado River and the electricity it generates for the western United States.

    Both the Brotherhood and the survivors in the mountain realize the jet stream is returning the fallout of the nuclear winter back to the region. Both must secure food and supplies to carry them through the return of the fallout. The survivors in the mountain repel attacks on the mountain by the Brotherhood and enter the next phase of nuclear winter prepared to battle the Islamic Brotherhood and drive them from the region.

GLOSSARY

- ADVERSE YAW: The tendency of an aircraft to yaw away from the applied aileron. Induced by rolling motion and aileron deflection, usually greatest at high angle of attack and full aileron deflection.
- AILERON ROLL: Rolling the aircraft around the longitudinal axis by used of the ailerons.
- AIRCRAFT AXES: There were three axes, which were mutually perpendicular and had a common point of intersection. The longitudinal axis parallel to the fuselage reference line, and aircraft rotation around this axis called roll or bank. The aircraft vertical axis was perpendicular to the longitudinal axis through the center of the aircraft, and rotation about that axis was yaw. The lateral axis was perpendicular to both other axes at the point of intersection. Rotation around that axis was pitch,
- ANGLE OF ATTACK (AOA): The angle between the chord line of the wing and the aircraft flight path (relative wind).
- ANGLE OFF: The angular measurement between the longitudinal axes of an aerial target and an attacking aircraft measured from the target's tail assuming the attacker was flying a pursuit attack.
- ASPECT ANGLE: The angle measured from the tail of an aircraft to define position relative to another aircraft regardless of the relationship of flight path.
- BARREL ROLL ATTACK: A maneuver used at high angles off and long ranges to arrive closer to the lethal envelope.
- BARREL ROLL (HIGH G): A maneuver used to cause an overshoot by an attacker who had maintained position in the lethal envelope during maximum performance turning. This last ditch maneuver leaves the defender with little or no maneuvering potential.
- BREAK: A maximum performance defensive turn into the attacker instantly destroying an attacker's tracking solution or position in the defender's lethal envelope.
- CLOSURE (Relative Velocity): The time rate of change of distance along the line of sight between aircraft.
- CONSTANT PEG Air Combat Training with MiG from 4477th TES, at TTR, early 1980s - 03/1988, follow-on to Have Idea
- DEFENSIVE SPLIT: A controlled separation of a defensive element in different planes used in an attempt to force the attackers to commit themselves to one of the defenders. A controlled separation with the defenders turning in the same relative direction but separated in the horizontal and vertical planes.
- DEFENSIVE TURN; The basic defensive maneuver designed to prevent an attacker from achieving a launch or firing position. A planned turn to prevent an attacking aircraft entering the defender's lethal envelope. Defensive turns can be from one g to maximum performance.
- DIVING SPIRAL: A near vertical accelerating dive using g and roll rate to destroy an attacker's tracking solution and gain lateral separation.
- DOM: Defensive combat maneuvering.
- ELEMENT: The basic fighting unit (two aircraft).
- ENERGY LEVEL (Es): Total energy state possessed for a given combination of altitude and airspeed (Each).
- ENERGY MANEUVERABILITY: A concept used to determine total in-flight performance

by measuring instantaneous and sustained maneuverability of an aircraft through its envelope.
- ENERGY RATE (ERs) A measure of the ability to gain or lose energy in terms of altitude and airspeed or combinations thereof.
- FME Foreign Material Exploitation
- HARD TURN (Single Direction Turn): A planned defensive turn in which by the angle-off, range, and closure of the attacking aircraft governs the intensity of the turn.
- HARMONIZATION: The adjustment of guns and sight of an aircraft, so when within effective range, the tracking index would indicate the impact point of the bullets.
- HAVE DRILL / HAVE FERRY Exploitation of 2 ex-Syrian MiG-17F from Israel, used for Air Combat Training at Groom Lake, USAF/USN joint project, predating Have Idea (1969)
- HAVE DOUGHNUT Exploitation of 1 MiG-21F-13, used for Air Combat Training at Groom Lake, USAF/USN joint project
- HAVE GLIB evaluation of foreign radar and threat systems. The systems were given names such as Mary, Kay, Susan, and Kathy and arranged to simulate a Soviet-style air defense complex. November 1970
- HAVE IDEA Evaluation MiG-21 and MiG-17F variants. May 1973
- HAVE PAD Exploitation of the Flogger MiG-23MS (1978)
- HIGH SPEED YO-YO: An offensive maneuver performed to maintain nose-tail separation and prevent the possibility of becoming engaged in a scissors maneuver.
- HI SPEED Yo-Yo. A maneuver utilized by an attacker, in the vertical and horizontal planes, to prevent an overshoot in the plane of the defender's turn.
- IN-TRAIL: Individual aircraft, one behind the other.
- JINKING MANEUVER: A series of rapid turn reversals or abrupt changes of roll/pitch angle at random intervals, to prevent an attacker from achieving a tracking solution. Usually employed with little load factor while gaining lateral separation.
- LAG PURSUIT ATTACK; An attack in which the nose of the attacker's aircraft remains pointed behind the defender's aircraft.
- LATERAL SEPARATION: Distance between an attacker and defender, measured in the horizontal plane perpendicular to the defender's flight path.
- LEAD PURSUIT ATTACK: An attack in which the nose of the attacker's aircraft remains pointed ahead of the defender's aircraft.
- LETHAL ENVELOPE: An area around every aircraft from which an attacking aircraft can launch a missile or achieve a gun tracking solution and expect to down the defending aircraft.
- LOW SPEED YO-YO: A maneuver employed to facilitate closure and at the same time allow an attacker to remain inside an opponent's turn radius. A maneuver to close on a target using a combination of altitude, airspeed, and cutoff.
- LUFBERV: A circular tail chase
- MANEUVERABILITY: The ability to change direction and/or magnitude of the velocity vector.
- MANEUVERING ENERGY. The ability to perform maneuvers because of energy possessed.
- MAXIMUM: Maximum afterburner power.
- MAXIMUM MILITARY POWER: Maximum non-afterburner power.
- MAXIMUM PERFORMANCE: The best possible performance without exceeding aircraft limitations.
- MAXIMUM PERFORMANCE MANEUVERING ENVELOPE: A maneuvering region for the wingman to achieve maximum performance maneuvers with optimum visual coverage

and mutual support.
- MAXIMUM RATE TURN: That turn at which the plane achieves maximum number of degrees per second.
- RHAW: Radar Homing and Warning
- SANDWICH: A maneuver designed to place the attacking aircraft/element(s) in train between the defending element(s).
- SEPARATION: Distance between an attacker and defender. Could be either lateral or longitudinal.
- SCISSORS: The execution of a defensive maneuver in which a series of turn reversals attempt to achieve the offensive after an overshoot by the attacker.
- SPLIT PLANE MANEUVER: Maneuver involving two aircraft in a defensive split, or four aircraft in a fluid separation to force attacker(s) to commit on one aircraft or element.
- TCA: Track Crossing Angle TOA-Angle-Off (Aspect Angle): The angle between the defender's line of flight and the attacker's line of sight measured in degrees (Track Crossing Angle).
- YF-110B MiG-21F-13 "FISHBED-C"
- YF-110C A MiG-21 "FISHBED" variant
- YF-110D A MiG-21 "FISHBED" variant
- YF-113A MiG-17F "FRESCO-C" used in HAVE DRILL program
- YF-113B MiG-23BN "Flogger-F"
- YF-113C MiG-17F (actually a Chinese-built J-5) "FRESCO-C" used in HAVE PRIVILEGE program
- YF-113E MiG-23MS "Flogger-E"
- YF-114C MiG-17F "FRESCO-C" (the one used in the HAVE FERRY program)
- YF-114D MiG-17PF "FRESCO-D
- V-N Diagram: A plot of load factor versus velocity used to provide a measure of instantaneous maneuverability.

List of Acronyms and Abbreviations
- afterburner   Afterburner
- ACM   Air Combat Maneuvering
- AERO1A   F-4B Missile Control System
- AFFTC   Air Force Flight Test Center
- AGL   Above Ground Level
- AIM-7E   SPARROW III Missile
- AIM-7E-2   SPARROW III (dogfight) Missile
- AIM-9B/D   SIDEWINDER Missile
- AMCS   Airborne Missile Control System
- AN/APR-25   Radar Homing and Warning System
- APG-59   F-4J Radar
- APQ-72   F-4B Radar
- APQ-94   F-8E Radar
- ATOLL   Soviet Air-to-Air Infrared Missile
- AWG-10   F-4J Missile Control System
- Bingo   Minimum fuel state required for safe return to base
- B/N   Bombardier/Navigator
- CNO   Chief of Naval Operations

- COMOPTEVFOR Commander Operational Test and Evaluation Force
- CRT Combat Rated Thrust
- cw Continuous Wave
- DIA Defense Intelligence Agency
- DRV Democratic Republic of Vietnam
- EGT Exhaust Gas Temperature
- EI Electronic Intelligence
- FFAR Folding Fin Aircraft Rocket
- FME Foreign Material Exploitation
- Free F-4 Tactical Wingman
- FTD Foreign Technology Division of Air Force Systems Command
- g Acceleration due to gravity
- HAVE DOUGHNUT Project Name
- HEI High Explosive Incendiary
- I Band Radio Frequency Energy (8,000 to 10,000 KC)
- IMN Indicated Mach Number
- IR Infrared Radiation
- KCAS Knots Calibrated Airspeed
- KIAS Knots Indicated Airspeed
- KTAS Knots True Airspeed
- MBC Main Beam Clutter
- MiG-21 MiG-21 F-13 (FISHBED C/E)
- MM Millimeter
- MRT Military Rated Thrust
- MSL Mean Sea Level
- NATC Naval Air Test Center
- NATOPS Naval Aviation Training and Operational Standardization Program
- NM Nautical Miles
- Oblique Loop An overhead maneuver performed just off the true vertical. An angle of bank is held to facilitate maintaining visual contact with a target in the rear hemisphere
- Padlock A lookout technique that required looking solely at the visually acquired target.
- PLM Pilot Lock-On Modification
- PPS Pulses Per Second
- PRF Pulse Repetition Frequency
- PD Pulse Doppler
- Pk Predicted Kill
- q Dynamic Pressure
- SEA Southeast Asia
- TAC Tactical Air Command
- TAC LEAD Tactical Lead
- TAC WING Tactical Wing
- TCA Track Crossing Angle
- UHF Ultra High Frequency (Radio)
- VA Fixed wing, heavier than air, attack airplanes

- Vc	Closing Velocity
- VF	Fixed wing, heavier than air, fighter airplanes
- VID	Visual Identification

Made in the USA
San Bernardino, CA
26 May 2014